How To Test
Your Dog's IQ

How To Test Your Dog's IQ

by Dr. Walter A. Luszki, Ed. D., &
Dr. Margaret B. Luszki, Ph. D.

TAB BOOKS Inc.
BLUE RIDGE SUMMIT, PA. 17214

FIRST EDITION

FIRST PRINTING—MAY 1980

Copyright © 1980 by TAB BOOKS Inc.

Printed in the United States of America

Reproduction or publication of the content in any manner, without express permission of the publisher, is prohibited. No liability is assumed with respect to the use of the information herein.

Library of Congress Cataloging in Publication Data
Luszki, Walter A.
　How to test your dog's IQ.

　　Includes bibliographical references and index.
　　1. Dogs—Psychology—Testing.
　　2. Animal intelligence—Testing.
I. Luszki, Margaret Elizabeth Butler Barron, 1907-　　joint author.
II. Title.
SF433.L87　　156'.3'932　　79-12606
ISBN 0-8306-9801-9
ISBN 0-8306-1143-6 pbk.

Contents

Introduction..7

1 What Is Human Intelligence? 13
Intellectual Functions—Intelligence Tests—Intelligence Scales—Fluid and Crystallized Intelligence

2 Measuring Human Intelligence..23
Standardized Tests—Stanford-Binet Test—Wechsler Scale

3 Intellectual Functions of Human Intelligence 31
Subtests—Verbal Subtests—Performance Subtests

4 Animal Intelligence and Its Measurement.......................45
Learning—Problem-Solving Ability—Reasoning or Inference—Ability to Remember—Ability to Form Concepts

5 Intellectual Functions of Dog Intelligence.......................57
Fund of Verbal Knowledge—Common Sense and Judgment—Attention—Concentration Ability—Abstraction Ability—Planning Ability—Ability to See Things as Wrong or Missing—Fine Eye-Muscle Coordination and Space Perception—Problem-Solving Ability—Ability to Adjust to a New Situation and Motivation

Effects of Deprivation and Enrichment 69
Human Infants vs. Puppies—Testing of Other Animals

7 Sources of Human and Dog Intelligence 89
Effects of the Environment—Inherited Characteristics—Critical Periods of Dog Intelligence

8 General Aspects of Test Administration 103
The Examiner—Test Sequence—Test Preparation—Time Limit—Scoring—Test Results—Obtaining DIQ (Dog Intelligence Quotient)

9 Administering and Scoring the Tests 113
Test Number 1—Test Number 2—Test Number 3—Test Number 4—Test Number 5—Test Number 6—Test Number 7—Test Number 8—Test Number 9—Test Number 10—Test Number 11—Test Number 12—Test Number 13—Test Number 14—Test Number 15—Test Number 16

10 Increasing Your Dog's IQ...147
Activities and Experiences—Supplements to Formal Training—Reinforcements

11 Differences in Intelligence Among Breeds 155
Breed Ranks—Results

Index..159

Introduction

We became interested in the intelligence of our German Shepherd dogs when we noticed that one seemed to be smarter than the other. At that time Rebel was the only dog able to use his paw to open the gate leading up to the front door. The gate was closed but not locked and he would place his paw at the lower bottom left and manipulate the opening of the gate. Later, some of the younger dogs seemed to learn from him by imitation.

Rebel was also able to open a sliding glass door and could push up a fork latch of a chain link fence gate. We felt that his problem-solving ability was of a higher level than that of the other dogs and that a battery of tests would help evaluate the relative intelligence of different dogs.

In considering intelligence tests for dogs, remember that breeds have been developed over many years to serve specific functions. Such dogs would show a high level of intelligence in some problem situations and appear stupid in others. The tests proposed in this book are designed to test puppies from the same litter or dogs of the same breed of about the same age.

Many times breeders are confronted with evaluating the merits of puppies in a particular litter. One aspect of such evaluation is intelligence. For those who are interested in dog training, intelligence is perhaps one of the most important factors. At best these test items are simply a rough guide to provide additional information to use in the selection process. Some dog owners may wish to use the test as merely a fun game. They may have no interest in comparing two dogs of the same litter or two dogs of the same breed of approximately the same age.

There have been recurring discussions of the intellectual abilities of dogs and particularly whether dogs can *think* or not. Various statements have been made by veterinarians and others that dogs cannot think.

The following was written some time ago in the *American Weekly*: "Can Dogs Think? Whatever You Think, This Veterinarian Answers NO" by Arthur Trayford, D.V.M. and Gladys Hall:

> Over the years we have heard many stories calculated to prove that dogs are thinking animals. But dogs do *not* think. Indeed they cannot, the cells of their cerebrum (thinking apparatus) being, very definitely, underdeveloped. It is always possible to explain "thinking dog" stories, to show that whatever happened required no thought.
>
> Among our patients, for example, is a beautiful Golden Retriever named Bobo. Every afternoon Bobo waits for the school bus. When the three children of the family emerge he settles down for a snooze.
>
> "But," says his mistress, "if one of the three is missing he covers the neighborhood, barking. He's even run the mile and a half to school, searching."
>
> The explanation here is very simple. Bobo is aware, through his sensory perception, that one of "the smells" is missing and his retrieving instinct takes over.
>
> Another patient of ours is a 9-year-old Boxer known as Maggie. Following an auto accident, Maggie's owner was confined to bed with both legs in casts. Later, with one leg in a walking cast, she was able to go up and down stairs. Maggie would precede her, pacing herself so she was one step ahead or one step below. One day, on the way downstairs, Maggie stopped. Her owner pushed, scolded and pleaded. But Maggie would not budge.
>
> "Finally," says her devoted owner, "I had sense enough to look down. On the step below Maggie was a pencil. Had I stepped on it, almost certainly I would have had a bad fall. I have my doctor's word that if I'd broken my ankle again it is doubtful I would be walking today."
>
> The explanation? Maggie could have smelled the wood or the lead in the pencil. Or, head down as she descended, she could have touch the "foreign object" with her whiskers. A dog's whiskers, like the feet pads, are tactile. Or she could have been stopped by something that had nothing to do with the pencil; a sound or even the adrenalin smell of the fear her owner felt as she negotiated the stairs.
>
> Dogs react uneasily to anything that is strange because it is with the familiar that they are secure. If a piece of furniture is pushed from its accustomed position or an owner deviates from the normal routine, a dog instantly will be disturbed.

It is because dogs react to the familiar as they do that the first law in their training is repetition. The German Shepherd, prepared from the time he is 6-months-old for the life of a Seeing Eye Dog, learns through the repetition of specific words—accompanied by specific rewards for obedience and specific punishments for disobedience—not to cross the street while traffic moves in front of him. For dogs, totally color-blind, cannot distinguish a red light.

Many dogs develop an associated response to the spoken word through the tone of voice in which that word is spoken. "Walk" evokes a happy response because it is said in a spirited tone and followed, usually, by the fun of going out; "down" an unhappy one because of the unusually severe tone of the command.

"But how," demands the owner of a certain Springer Spaniel, "can my Springer tell the time if he can't think? Every evening at 6:45 he dashes for our back door and stands there, tail wagging, when it is just about the time that my husband's car pulls into the garage."

The Springer, through his acute sense of hearing, may well pick up the sound of his master's car long before it can be heard by any human ear. It's also probable that cooking preparations and odors habitually precede the car's arrival—and the association of sound and smell, repeated night after night, is what alerts the Springer.

Strange tales indeed are whispered about dogs that "know" when there's to be a death in the family, a foreknowledge sometimes exhibited by a continuous howling. Dogs, like certain primitive people, may well have a clarivoyant power. But often, we are sure, those that "howl like banshees" are simply put out by the indifference and neglect shown them at such times.

Dogs are "dumb animals" only in that they can neither think nor speak. They can perceive and understand what goes on around them. And since dogs, like all living things, are individual, some have greater perception than others. Moreover—from the time dogs are 1-year-old they have a keen awareness for their environment and the people with whom they share it. It is this, undoubtedly, that has made them such satisfying and lovable companions down through the years.

After reading this widely circulated article there is no doubt that many dog owners felt they had a dumb animal on their hands. The question of whether a dog can think cannot be answered by a categorical yes or no. To give the animal a fair chance we ought to first define what we mean by thinking and then reconsider the veterinarian's opinion.

The psychologist's bible, *A Comprehensive Dictionary of Psychological and Psychoanalytical Terms* (English and English) defines *thinking* as:

- Any process or activity not predominantly perceptual by which one apprehends an object or some aspect of an object or situation. Judging, abstracting, conceiving, reasoning, imagining, remembering and anticipating are forms of thinking. Although thinking is thus negatively defined by reference to perceiving, the two processes are not antagonistic but supplemental. Either may merely predominate in any given cognitive process.
- Problem-solving that involves primarily ideas rather than perceiving and overt manipulation.
- Meditating or reflecting upon a problem in order to understand the relationships involved.

Further, *apprehension* is defined as the awareness of a relatively simple object; it involves a greater ideational content than is usually included in perceiving, yet it does not quite reach the level of thorough understanding meant by comprehension. As contrasted with judgment, apprehension is direct and immediate, judgment is mediate. One perceives a shout, apprehends a danger, judges that it is unwise to proceed. The verb apprehend is limited to this meaning.

When viewed in light of these definitions, the dog does have in his intellectual makeup some aspects of thinking. The experiments on animals including dogs mentioned in this book as well as the proposed dog intelligence test will clearly show that dogs can remember—one form of thinking. There is some indication that they can anticipate, reason and perhaps even abstract. They have been known to solve problems. Problem solving is the chief intellectual function measured by our test. One of the test items measures attention, a mental function described as taking things in through the senses and registering them in the brain automatically without any effort on the part of the dog. Attention includes more than perceiving and it comes closer to the ideational content of understanding than does perceiving. Some animal researchers say that dogs have insight. *Insight* is the process by which the meaning, significance, pattern or use of an object or situation becomes clear. The dog learns different parts of the apparatus on different occasions and later makes use of the separate experiences to solve a new problem.

Because of the extensive work that has been done on human intelligence the senior author started with the intellectual functions identified by Wechsler (1949, 1958). These intellectual functions include:

- general information
- common sense and judgment
- abstraction ability
- concentration and arithmetical reasoning
- fund of verbal knowledge
- attention
- ability to see things wrong or missing
- planning ability
- ability in spatial perception
- manipulative ability
- fine eye-hand coordination ability
- problem-solving ability
- motivation
- ability to adjust to a new situation

Five of these seem applicable to dogs and they provide the basis for the proposed IQ Test. These are:

—Fund of receptive verbal knowledge (WAIS-WISC: Vocabulary)*
—Common sense and judgment (WAIS-WISC: Comprehension)
—Concentration ability (WAIS-WISC: Comprehension)
—Attention (WAIS-WISC: Digit Span)
—Problem-solving ability (WICS: Mazes)

*WAIS: Wechsler Adult Intelligence Scale
WISC: Wechsler Intelligence Scale for Children

Chapter 8 deals with the general aspects of the administration of the test. The tests themselves and their administration are contained in chapter 9. To form a better basis for understanding The L & L Dog IQ Test, chapters 1, 2, and 3 deal primarily with human intelligence followed by a comparison of human and animal intelligence, intellectual functions of dog intelligence and sources of human and dog intelligence.

Dr. Walter A. Luszki and Dr. Margaret B. Luszki

Chapter 1
What Is Human Intelligence?

The term *intelligence* when applied to animals (the word animals will be used to mean the non-human ones, including dogs Fig. 1-1) has been used so vaguely and in so many different ways that some people feel it would be better to avoid using the term. When talking about human intelligence, we have a reasonable agreement on the meaning of intelligence in large part because of the tests used to measure intelligence.

It is important for the reader to understand the basic principles of human intelligence before delving into the "intelligence" of dogs. By becoming familiar with human IQ testing, the reader will be able to better test his dog's IQ.

Most of us have some idea of what intelligence is. Among people we know we classify certain ones as intellectually brilliant, others as intelligent and still others as dull or not too bright. Webster's dictionary defines the term:

> the capacity to know or comprehend; the capacity for knowledge and understanding, especially as applied to the handling of novel situations; the power of meeting a novel situation successfully by adjusting one's behavior to the total situation; the ability to apprehend the interrelationships of presented facts in such a way as to guide action toward a desired goal; the quality or trait of understanding; mental acuteness; sagacity; knowledge; particular or general information.

Synonyms of intelligence are mental alertness, cleverness, understanding, knowledge, smartness and comprehension. Antonyms include dullness, mental retardation, slowness and stupidity.

There is more agreement on the behaviors referred to by the term intelligence than there is on how to interpret or categorize them. Three concepts recur frequently in attempts to start its connotations: ability to deal effectively with tasks involving abstractions; ability to learn and ability to deal with new situations.

The following is a widely accepted definition of intelligence:

> The ability to undertake and carry through activities that are characterized by difficulty, complexity, abstraction and adaptiveness to a goal.

The following two definitions limit themselves to stating operations by which intelligence is to be distinguished from other constructs:

- That hypothetical construct which is measured by a properly standardized intelligence test. Intelligence tests can be—and in fact have been—devised and standardized without having any particular or clear definition of intelligence.
- The individual's total repertory of those problem-solving and cognitive-discrimination responses that are usual and expected at a given age level and in the large population unit to which he belongs. The *usual and expected response* has been defined, by implication of test standardization, as one of which 65 percent to 75 percent of the given population are capable. What is thus usual and expected changes qualitatively as well as quantitatively with age and with the population. Intelligence tests, regarded as samplings of the total repertory, must reflect these changes. The intelligence level is measured by the proportion of the responses, usual and expected in the population, that an individual manifests in a standarized sample of task-demand situations.

INTELLECTUAL FUNCTIONS

Intelligence is not a function of any set of bodily structures, such as the nervous system, sense organs, and receptors, but of the whole organism. When animals at different levels of evolution are studied, it is noted that behavior at one level differs from that at another. Under comparable circumstances the behavior of worms, for example, is not as flexible nor as versatile as that of the behavior of rats in response to problems in their environment. We say, therefore, that rats are smarter or more intelligent than worms. In like manner we say that man's behavior is more intelligent than that of monkeys.

Psychologists accept the comment that intelligence is not a unique entity, but a combination of various traits, skills and abilities which serve certain goals. When we recognize or identify a piece of behavior as either smart or stupid, we are talking intelligence.

A number of scholars during the past century have made significant contributions to the understanding of the concept of intelligence. The search for more information continues. It seems to be an unending search. At present it is agreed by most experts that intelligence consists of a group of distinct but interrelated mental qualities which vary from person to person. Viewpoints vary as to the number and identity of these separate mental functions, and the ways and extent of their interconnectedness.

INTELLIGENCE TESTS

One of the first scholars to study those intellectual functions whose sum total was considered to consist of intelligence was the French psychologist Alfred Binet. In 1895 Binet wrote that the general mental capacity of persons could be examined by administering a series of tests of attentiveness, imagination, mental imagery, memory, suggestibility, aesthetic appreciation, moral sensibility, mechanical and verbal comprehension, the capacity to sustain muscular effort and visual judgment of distance.

One of Binet's greatest contributions was his statement about the selection of tests. He wrote that it made little difference what kind of test items were used, as long as that task in some way was a measure of the person's general intelligence. This accounts in part for the large variety of tests used in the original Binet scale. His assumption also accounts for the fact that certain kinds of test items were useful at one age level but were not used at a different age level.

Binet and Simon (1916) used one score in their intelligence test. This score was called the *mental age*. They included a great variety of tasks in their test. Only those tasks that a child might be expected to pass were used.

As early as 1904, about a decade after Binet, Charles Spearman (1927) in England identified two factors in intelligence: a general factor which he designated as g and a specific factor which he called s. He found that most of the mental tests then available correlated positively but that the correlations were not as high as their reliability would have made possible if they were all measuring the same thing. He, therefore, concluded that each test must be measuring two factors: general and specific factors. He defined the general factor as general intelligence. The specific factor was unique to each test. He suggested that every intellectual test must measure individual differences in a universal ability which he called g and also differences in a specific ability, s.

E.L. Thorndike (1927) in 1909 created much confusion among psychologists when he disagreed with Spearman on what consitututes intelligence. Spearman used certain statistical techniques to support his conviction that a general intelligence factor was present. Thorndike, using the same data but different statistical procedures, argued that there was no evidence for a general ability factor. Thorndike maintained that the studies clearly showed that intelligence consisted only of specific factors.

The 1931 L. L. Thurstone (1938) presented the idea that intelligence consists of several independent skills or abilities, which he termed *primary mental abilities*. He devised a measurement technique designated as the method of multiple factor analysis. He gave 56 tests to students at the University of Chicago and found six key factors: Verbal *(V)*, Number *(N)*, Spatial *(S)*, Word fluency *(W)*, Memory *(M)* and Reasoning *(R)*. The verbal factor V is located in vocabulary tests and in tests of comprehension and reasoning. The number factor N is found in simple arithmetic tests. The spatial factor S is concerned with visual form relationships. The memorizing factor M is found in tests which include rapid rate learning, including memory for words, digits and designs. The word fluency factor W deals with the ability to think of words rapidly, as in anagrams and rhyming. In subsequent experiments he identified another factor: *perceptual ability*. He developed seven tests, each aimed at measuring one of the seven factors. If Thurstone had succeeded in breaking intelligence down into seven distinct factors, scores from these tests would not be correlated with each other. But when he gave the tests to new subjects and determined the corelation coefficients between the tests, he discovered that the tests in fact were correlated. Thus, he was unable to develop tests where each only measured unique components. His research could mean that in addition to the special factors which he called primary mental abilities, there was a general intelligence factor.

Thurstone's primary mental abilities list has been considered by some as a list of the basic elements of intellectual functioning. Some persons have compared his list to the chemist's list of elements. Many investigators found similar factors, making it seem that Thurstone had been getting to the basic elements of the human mind. Other investigators disagreed with him and devised tests whose items seemed to involve many other types of mental processes.

INTELLIGENCE SCALES

David Wechsler (1958) devised the most widely used intelligence scales, one for children and another for adults. They are termed the Wechsler Intelligence Scale for Children (Revised) and Wechsler Adult Intelligence Scale. He defined intelligence as the "aggregate or global capacity of the individual to act purposefully, to think rationally and to deal effectively with his environment." He calls intelligence global because it typifies the person's behavior as a whole. Because it is composed of elements or abilities which can be differentiated, it is aggregate. To him intelligence is more than the sum of mental abilities. How these abilities are combined with one another or their configuration is an aspect of intelligence. Wechsler also postulated that different kinds of intelligent behavior may require different degrees of intellectual ability, but an excess of any ability may add relatively little to the effectiveness of the behavior as a whole. For example, to behave intelligently, one must have a reasonably good retentive memory. It may not help much in handling life situations effectively to have an overabundance of this skill. A high school student may score in the 99th percentile in abstraction ability on the WAIS, but such a high level of this ability may not improve the student's coping skills. W. A. Luszki (1966) reported the test performance and behavior of a 32-year-old deaf-retarded male. The retardate provides a striking example of superior functioning on the WAIS Block Design subtest (fine eye-hand motor coordination skill), but poor adjustment to the real world as we know it.

Wechsler noted that there are many definitions of intelligence which are too restricted or too limited. Some people emphasize the ability to reason or to conceptualize. Others hold up common sense and judgment as the all important trait of intelligence. Other people consider that a fund of verbal knowledge and general information are the two abilities which alone constitute intelligence. Still others think that the ability to solve complex problems is the key ingredient in intelligence.

All of these are certainly aspects of mental functioning and one cannot deny that they are part of intelligence, but these definitions are too incomplete. They are concerned with only a limited range of skills and abilities that can be subsumed under the heading of intelligent behavior. Intelligence may manifest itself in each of these ways, but it may also be seen in many other ways. It may manifest itself in non-intellective factors. Wechsler included the following components under this heading: drive, persistence, motivation and goal awareness. He agreed with W.A. Alexander (1935), who identified what he called factors X and Z. These pertained not to abilities as ordinarily understood but to aspects of behavior which we know as temperament. They concern such traits as drive, persistence and interest. In certain kinds of achievements Alexander found that these traits were important. For example, Alexander found that for success in science the verbal factor accounted for 31 percent of the success, but X (temperament factor) contributed 55 percent toward success in science. In contrast, on the Kohs block test where the task is to copy a geometric design by arranging small multicolored cubes, the temperament factor did not contribute anything to success. Wechsler was convinced that the temperament factors form part and parcel of what is finally necessary for intelligence behavior. He called these the non-intellective factors in general intelligence.

The non-intellective factors of intelligence include all affective and connotative skills which enter into global behavior in any way. A child may do poorly on an intelligence test because he is not interested or upset emotionally. Here his emotional behavior is specific to the situation. He might do much better if retested. On the other hand, if he does poorly on the test because he is basically impulsive or emotionally unstable, his behavior represents a permanent mode of response. We would not expect much improvement in test performance if we retested him.

Ronald Nygad (1977), a Norwegian researcher, studied the concept of persistence of behavior which has been a central concern of goal-directed theories of motivation. He also studied the relationship of persistence to intelligence. His theory is that an individual will persist at any activity as long as total motivation to perform it is stronger than the total motivation to perform an alternative activity. To account for persistence in an achievement situation, one must consider the relative strength of a person's motive to achieve success, his motive to avoid failure, the perceived difficulty for working at the task (e.g., parental demands). The attractiveness of the alternative to which the person can turn if he or she abandons the initial task must also be considered.

A few months ago the authors worked with a 16-year-old female patient who had Borderline Intelligence (WAIS Verbal Scale IQ: 72), just three IQ points above the mental retardation level of 69. Yet, she was in the ninth grade and able to read and do arithmetic at the eighth grade level. To keep up with her homework she studied from five to six hours daily. In our opinion her achievement was accounted for almost entirely by persistence. Her mother, retired on disability and confined to bed, supervised her homework. There were many personality problems that this adolescent developed because of her rigorous study habits, but she was passing academically.

FLUID AND CRYSTALLIZED INTELLIGENCE

Besides the scholars mentioned above who contributed immensely to the understanding of the nature and definition of intelligence, there were many others. Of this latter group at least three should be singled out. J. P. Guilford

Fig. 1-1. When applied to dogs, the term intelligence has been used so vaguely and in so many different ways that some people feel it would be better to avoid using the term. (Photo courtesy of Allied Food Inc., Wayne Dog Food.)

(1967) together with R. B. Cattell (1963) are two psychology theorists who developed a variety of tests of different intellectual functions. Over the past three decades each has developed his theory. Through a Herculean task Guilford and his students have identified tests for 98 out of 120 different abilities. They have also shown the intercorrelations between these abilities. Cattell has developed the concepts of a *fluid intelligence* and a *crystallized intelligence*. Cattell hoped to measure fluid intelligence by tests that were *culture free* or at least *culture fair* for all persons regardless of the richness or deprivation of their educational and cultural background. Cattell's second form of intelligence, crystallized, is sensitive to each person's unique educational, environment and cultural background. The Swiss psychologist, J. Piaget (1947), postulated qualitative elements in the development and utilization of intelligence. To Paiget intelligence is thought of as an aspect of biological adaptation. It is coping with the environment and organizing thought and action in different ways as the person grows and development takes place. Intellectual abilities do not emerge and then deteriorate as the person gets older. Instead, the intellectual components improve as one grows as a function of related intellectual experiences.

Chapter 2
Measuring Human Intelligence

There are many individual and group tests available which have been standardized to determine human intellectual capacities. Generally, these instruments sample such traits as comprehension, vocabulary, facility of dealing with spatial relationships, efficiency of attention and concentration, memory, speed of visual perception, psychometer coordination and rate and quality of reasoning ability.

In clinical practice, it is desirable whenever possible to utilize an individually administered scale, since motivation and behavior can be evaluated more adequately. Also, an individually conducted examination provides a more valid basis for making a sound estimate of intellectual potential and function.

There are numerous other mental ability tests sampling verbal and nonverbal capacities which may be employed individually or in group administration. Some are specially adapted to individuals who are not fully literate in English, have speech or auditory defects or who are visually handicapped.

STANDARDIZED TESTS

The range of standardized tests extends to all age levels. The Revised Stanford-Binet Intelligence Scale, Form L-M; California Test of Mental Maturity; and Otis Self-Administering Tests of Mental Ability may be cited as examples of clinically useful items.

The Wechsler Adult Intelligence Scale (WAIS) and the Wechsler Intelligence Scale for Children-Revised (WICS-R) are particularly effective instruments for use with adults and children.

James McKeen Cattell first used the term *mental test* in 1890. The mental tests at that time were chiefly experiments involving specific sensory, motor, perceptual and memory tasks. In 1908 Alfred Binet and his colleagues published a comprehensive intelligence test. He tried to find out how *smart* and *stupid* children differed. He used a large number of test items including recall of digits, size of head, moral judgment, tactile discrimination, addition, graphology, suggestibility and many others. He eventually defined intelligence as "the tendency to take and maintain a definite direction; the capacity to make adaptations for the purpose of attaining a desired end; and the power of auto-criticism."

The school officials in Paris needed a tool to segregate the child who was bright from the mentally retarded child. They requested Binet to devise a method for identifying children of low intelligence. The test Binet developed was welcomed in America as a research technique and as a way of studying children who were not intellectually normal. In 1916 Lewis M. Terman produced the Standford Revision of the Binet Scale. This revision included tests for normal and superior children as well as retarded children. In 1937 Terman and Merrill published Forms L and M of the Standford-Binet. In 1960 another revision came out— embodying the best of the 1937 revision into a single Form L-M and improving the scoring system.

One of Binet's contributions was to replace the idea of separate functions with the concept of general intelligence. Starting with the notion that some children were smart and others dull, he also noted that those who were best on tests of judgment were also better in vocabulary, attention, memory and other functions. This showed him that the tests were correlated, and this in turn demonstrated some underlying unity among these tests of intelligence. General mental ability referred to the characteristic that accounts for the relationship among the tests.

Binet had to do a lot of trial and error work to include or exclude various test items. If color matching did not correlate with other items of intellectual ability, it must not be influenced by the common factor. If knowing certain words correlates with the tests of abstraction, both must measure intelligence.

In the selection of test items Binet recognized the increase of mental ability with age. The older child is superior in following instructions, evaluating his own ideas and in understanding the meaning of words. He concluded that a good mental-test item should be easier for older children than for younger ones. An item should not be used if only 10 per cent of children of any age can pass it, for such an item is too difficult, not discriminatory and does not reflect mental growth. Binet preferred items on which success is related to age. He also assumed that it was important to measure a general ability running through all tasks. A good test item, he felt, should correlate with the remainder of the scale. Binet located items in the scale according to their difficulty for children at each age. A test which about 60 per cent of children of a particular age can pass is placed at that particular age level.

STANFORD-BINET TEST

The Stanford-Binet Test consists of items that measure the intellectual capacity at each age level. At age 10, for example, a child of average intelligence can:

- Repeat six digits forward.
- Find absurdity in a simple story.
- Define *pity* and *grief*.
- Give two reasons why most people would rather have an automobile than a bicycle.

The experienced examiner first tries items for a mental level below that expected. In determining where to begin one must take into account the chronological age, grade placement, general behavior in the test situation and any other pertinent information that may be available. The examiner determines the basal age, which is the level at which the child passes all the items. It sometimes happens that a person passes all of the tests at a higher age level than the one in which his first failure occurs. This does not change the base from which the score is determined. At each age level in the L-M scale, an extra test has been included for use as a substitute when a test has been spoiled in giving. An alternative test may not be given for a test which has been failed.

Scoring

In Form L-M the range of mental development is from age 2 to Superior Adult III. From ages 2 to 5 there are six tests at each half-year. Above age 5 the tests are spaced one year apart. Above age 14, the levels are even wider apart.

The subtests involve verbal and nonverbal items. Tasks concerned with objects and pictures are used at younger ages. At the upper end of the test there is more reliance on verbal problems and on abstract thinking. An attempt is made at objectivity in scoring by means of a scoring guide which contains examples of acceptable and unacceptable answers.

Mental Age

The scoring system initially involves the determination of a mental age. This mental age is the chronological age at which the average child does as well as the subject does. Bill, let us say, is 7 years old. On the test items he does as well as the average child of 9. Thus, he earns a mental age of 9 on Form L-M. The mental age is determined by crediting the subject with his basal age plus all additional credits earned beyond his basal. At the lower end of the scale, tests are grouped at half-year intervals including years 2, 2½, 3, 3½, 4, 4½ and 5. Each test passed in this age group earns a credit of one month toward the mental age. From year 6 through 14 each group represents an interval of 12 months. Accordingly, we count two months toward the mental age for each of the six tests at these levels.

The following is an example of determining the mental age. Jim is 7 years old. He passed all tests at the six year level so his basal level is VI. He receives a base credit of six years of mental age. He passed four tests passed at Year VII, so we add eight months credit; four at VIII adds eight months credit; two at IX adds four months credit; and one at X adds two months credit. He failed all tests at XI. Thus, he earned a mental age of 7 years and 10 months.

The mental age determines the subject's performance. If his mental age is greater than his chronological age then he is a bright child. After age 15 or 16 the mental age growth is slow. The average 22-year-old has a mental age well below

Table 2-1. Percentage Distribution of IQs.

IQ	PER CENT	CLASSIFICATION
160-169	0.03	Very superior
150-159	0.2	
140-149	1.1	
130-139	3.1	Superior
120-129	8.2	
110-119	18.1	High average
100-109	23.5	Normal or average
90-99	23.	
80-89	14.5	Low average
70-79	5.6	Borderline defective
60-69	2.0	Mildly retarded
50-59	0.4	Moderately retarded
40-49	0.2	Severely retarded
30-39	0.03	Mentally defective

20. Thus, the mental-age units used for higher ages are not directly related to the average performance at these ages.

IQ

The mental age score is converted into an IQ by referring to tables in the Stanford-Binet manual. The IQ is an indication of the subject's relation to others in his age group. The IQ was originally introduced as a quotient representing the subject's degree of mental development. The mental age was divided by chronological age and multiplied by 100, as follows: IQ = MA/CA × 100.

The purpose of multiplying by 100 is merely to eliminate decimals. Thus, if an individual's MA and CA are the same, he is of average intelligence and his IQ is 100. If his MA is 10 and his CA is only 8, his IQ will be 125. IQ's for the L-M Form are found by looking at the Pinneau Revised IQ Tables (Manual for the Third Revision, Form L-M, Stanford-Binet Intelligence Scale). One finds the MA in years and months and looks opposite the appropriate CA and reads the IQ score.

It is customary to refer to levels of intelligence in terms of IQ ranges as indicated in Table 2-1. About 46 per cent of the cases making up this sample fall between IQs of 90 and 109. This corresponds in a general way to what we call average intelligence.

What is a person like who has an IQ of 25, 50, 65, 140? Those who have an IQ of up to 25 are profoundly retarded. They never learn to avoid the common dangers and would soon die if someone else did not care for them. Many of them never learn to dress themselves or to say simple words. Some never learn to sit up, and they stay in bed all their lives. Those with an IQ from about 26 to 50 may be able to do simple work under close supervision, such as some janitorial duties or digging a ditch. They may learn to talk some. Those with an IQ from about 55 to 69 are mildly retarded. They can learn to read and write at a low level and can do certain types of routine factory work. As adults they have the mental capacity of average 7-to-12-year-old persons. They cannot be expected to go beyond the fifth or sixth grade in school.

A follow up study was made of children whose IQ was 140 or more. It was found that as adults these gifted individuals had lower suicide and insanity rates than the general population. Ninety percent entered college and of these 93 percent graduated. The average earnings of this group far excelled those of the general population. These gifted persons made great achievements in books and articles published, college degrees received and professional levels attained.

WECHSLER SCALE

While one of the important applications of the Wechsler Scale is in the area of diagnosis, the primary function of the Scale is to establish the overall intelligence level of the subjects tested. This is furnished by the IQ which a subject attains on the scale. Therefore, it is important to keep clearly in mind what an IQ stands for. An IQ on the Wechsler Scale (as in most scales) is a measure of relative position. It indicates how an individual compares with other individuals of like age with respect to scores attained on the tests. The obtained IQ is an arbitrary figure dependent on the way it has been calculated and the range of the test scores of the population examined. By convention the mean IQ has been set at 100 and other demarcations are expressed in terms of measures of deviation from this mean.

The IQ classification for the Wechsler Scale is in substantial agreement with IQ tables for other scales except at the extreme limits. Compared with other intelligence scales (and in particular, the Revised Stanford-Binet), one may expect the Wechsler Scale to give higher IQs at the lower levels (IQs 75 and below) and lower IQs at high levels (IQs 120 and over).

Today the most important intelligence tests are those included in Wechsler's intelligence scales. The Wechsler

Intelligence Scale for Children (WISC) was devised in 1949 and tests children from 5 to 15. In addition to yielding an IQ score, the test generally enables clinical psychologists to obtain diagnostic impressions that aid in arriving at a qualitative picture of the personality. A study of testing programs in institutions for the mentally retarded has indicated that the WISC is surpassed in usage only by the Stanford-Binet. The WISC was revised in 1974. Revision of the WISC began with a call for critical comments from experienced psychologists. Some of the older items were dropped out and a number of new ones added. WISC-R emphasizes the global entity of intelligence. In so doing it probes into intelligence in many different ways through many different kinds of tests.

The Wechsler Adult Intelligence Scale (WAIS) supersedes the Wechsler-Bellevue and is a widely used test for measuring adult intelligence (ages 16 to 75 and over). As in the Stanford-Binet Test, questions are arranged in ascending difficulty. The subtests on both the WISC and WAIS are scored on the basis of speed, accuracy, comprehension and ability to discriminate and analyze, according to the nature of the task. The points accumulated for each subtest are weighted and totaled to obtain the Full Scale Intelligence Quotient. Verbal and Performance Scale IQs are separately computed, but these are always considered in relation to the Full Scale IQ.

Among individual tests the Wechsler and the Stanford-Binet are equally prominent with no other serious competitor. There are, however, a number of little-used tests of good quality. These include Leiter International Performance Scale, Merrill-Palmer Scale, Minnesota Preschool Scale, Draw-A-Man Test, Columbia Mental Maturity Scale, Valentine Intelligence Tests for Children, Point Scale of Performance Tests and Pintner-Patterson Scale of Performance.

There are a number of tests of infant development. The main aim of most infant tests is to determine whether the infant is showing development normal for his age rather than to assess mental level. The following are the more important infant tests: Cattell Infant Intelligence Scale, Gesell Development Schedules, Griffiths Mental Development Scale, and California First-Year Mental Scale.

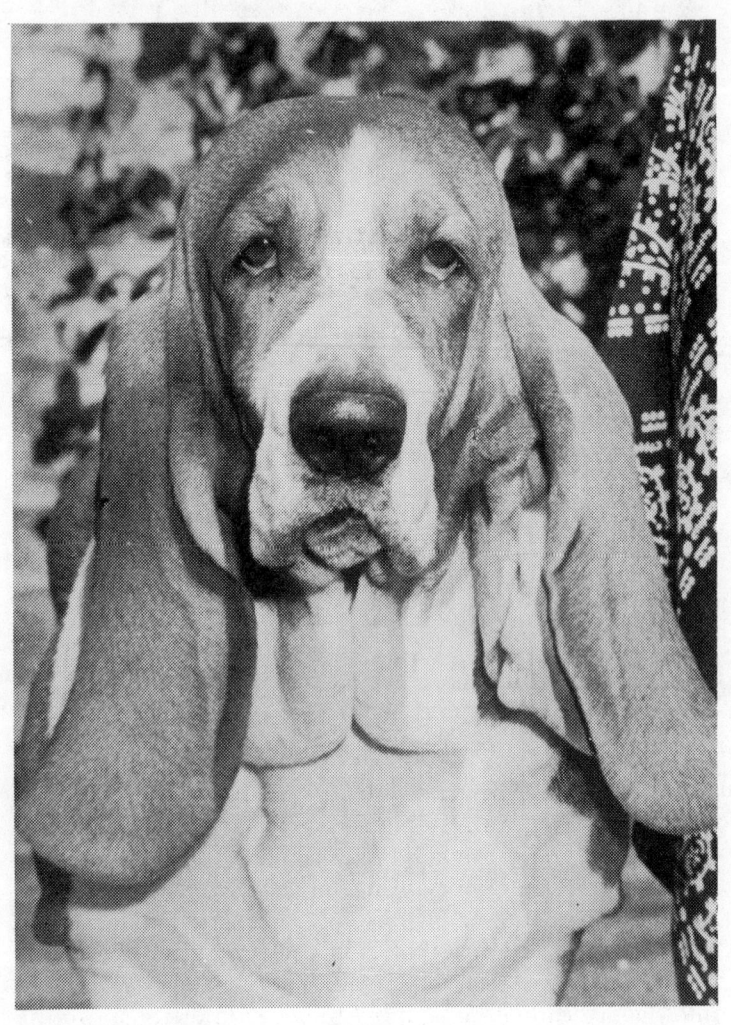

Chapter 3
Intellectual Functions of Human Intelligence

Intelligence can be looked upon as being composed of several functions or activities. David Wechsler organized a group of functions which he considered were the most important components of human intelligence. He developed his test in the Bellevue Hospital in New York City where he had to test persons who were mentally retarded, psychotic, illiterate or otherwise mentally disabled to help in determining appropriate plans for them. His test was of great value in military hospitals in World War II, and it was one of the most important diagnostic tools for clinical psychologists after that war.

The underlying theory of the Wechsler Scales is that the purpose of a general intelligence test is not to forecast how an individual will function in a given specific situation, for example, pass or failing a high school entrance examination, but how, on the average he may be expected to function in any and all situations calling for intelligent behavior. It assumes that intelligence is an interaction and not an entity. It also assumes that it is a part or facet of a larger whole, namely personality itself. This means that general intelligence cannot be equated to sheer intellectual ability but must be thought of as involving other capacities. The latter are what Wechsler called the non-intellective factors of intelligence. These include such traits as drive, persistence, perserverance and in some instances, aspects of temperament that pertain to interest and achievement. These non-intellective factors are not measured directly by the Wechsler Scales but may be inferred to some extent from the quality of the subject's responses.

SUBTESTS

The WAIS is composed of 11 subtests. The WISC consists of 12 subtests. Wechsler (1958) groups these sets of subtests in two series: verbal, which yields a *Verbal IQ* and performance, which yields a *Performance IQ*. The Verbal Scale includes subtests designated as information, comprehension, similarities, arithmetic, and vocabulary. The WAIS includes digit span as a regular verbal subtest. On the WISC, digit span and mazes are supplementary or alternate tests. The Performance Scale includes picture arrangement, picture completion, block design, object assembly and digit symbol (WAIS) or coding (WISC).

The grouping of the subtests into verbal and performance, while intending to emphasize a dichotomy in regards to possible types of abilities called for by the individual tests, does not imply that these are the only abilities involved in the tests. It does not imply that there are different kinds of intelligence such as verbal, manipulative, social, etc. It merely implies that there are different ways in which intelligence may manifest itself. The subtests are different measures of intelligence and not measures of different kinds of intelligence. They merely represent different areas of functioning. The dichotomy verbal and performance are broad groupings of these areas.

VERBAL SUBTESTS

The verbal subtests offer measures of intelligence by evaluating general information, comprehension, digit span, arithemetic, similarities and vocabulary.

General Information

Sample item: How many weeks are there in a year?

The test measures mainly how much you know about science, history, social studies and geography. This subtest is essentially a measure of the subject's fund of general knowledge and his awareness of the cultural world about him. Though his knowledge is dependent upon education and immediate background, it nevertheless calls for an ability to assimilate facts and information relative to one's environment. Types of items passed or failed often indicate the subject's range as well as area of special interests. When these are scattered and insufficient, it may indicate a lack of interest in mundane affairs or even withdrawal tendencies. This test presupposes a normal or average opportunity to receive verbal information; success does depend in a large degree upon the subject's educational and cultural opportunities. Persons of foreign tongue do poorly frequently on tests of information.

Comprehension

Sample item: Why should we keep away from bad company?

This subtest may be termed a test of common sense and judgment. That is, it measures the person's ability to make use of his intellectual assets in a manner not only logically correct but also relevant. Success on the test seemingly depends on possession of a certain amount of practical information and a general ability to evaluate past experience. It has been related to what has been referred to as *reality testing*. It is the appropriate understanding of and reaction to reality situations. The content of responses often reveals the subject's personal problems or preoccupations and personal attitudes.

Digit Span

Sample item: Say, "I am going to say some numbers. Listen carefully, and when I am through say them right after me, 8-4-7-2-9."

Digit span is a measure of rote memory and span of attention. Also involved to some degree is the subject's spontaneous organizational ability and freedom from perseverative interference. The ability to repeat digits backwards is an indication of flexibility and mental control. Successive failure in repetition of digits is often influenced by the subject's momentary tensions and hence is an indicator of anxiety.

Low scores on this test may be associated with organic defect, anxiety or inattention. Ordinarily an adult who cannot repeat four or five digits forward is either organically damaged or mentally retarded. Nevertheless, some mental retardates do well on this test. The test measures how well your brain takes in information automatically through the senses with relatively no effort on your part. If two persons sit in a waiting room for some 5 minutes and at the end of that time you ask both of them to come into your office and give each a piece of paper and pencil and ask them to write down all the things they saw in the waiting room, the person who listed the larger number of items (lamp, magazines, blue rug, piano, etc.) would have a higher level of attention than the other one. This test is not a too valid test of memory but rather of voluntary attention. Perhaps, present memory is measured to some degree by the digit span type of test.

Arithmetic

Sample item: A man has $14 and he spends $3.50. How much does he have left?

This test measures primarily arithmetical reasoning and facility with numbers. It is also a measure of concentration and attention and to some extent for capacity for sustained effort. It is not completely dependent on special numerical ability, but is a test of mental alertness. It is, however, dependent to a considerable degree on the amount of the subject's formal schooling. The subject must summon up appropriate knowledge for each problem in a limited time. Concentration involves a focusing of consciousness upon the current topic and the exclusion of other emotional or thought content. Arithmetic tests are usually devised to avoid verablization or reading difficulties. Clerks, engineers and businessmen usually do well on arithmetic tests, hence indicating that arithmetic tests reflect vocational and cultural interests and background. The ability to solve arithmetical problems has long been recognized as an indication of mental alertness. Arithmetical tests correlate highly with global measures of intelligence. The computational skills required to solve most of the problems are not beyond those taught in the elementary school or what the average adult could learn by himself. It has been noted that children who do poorly on this test often have trouble with other subjects. It appears that the combined scores of information, vocabulary and arithmetic subtests provide a fairly accurate estimate of the subject's school achievement.

The concept of attention and concentration are somewhat overlapping and have somewhat different meanings for different people. For purposes of this book they are defined as they are in *A Comprehensive Dictionary of Psychological and Psychoanalytical Terms* (H. B. and A. C. English). *Concentration* is defined as, "exclusive and persistent attention to a limited object or aspect of an object." *Attention* is defined as, "the active selection of, and emphasis on, one component of a complex experience, and the narrowing of the range of objects to which the organism is responding."

What is the difference between attention and concentration? First there can be no concentration without attention. Also, concentration involves a generally longer time span than attention. Memory is required in both attention and concentration, but generally to a larger extent in concentration. In any mental activity attention precedes concentration and continues during the concentration activity. Internal and external distractions interfere with concentration even to the point of eliminating it in the mental process. When a decreasing amount of concentration occurs, attention is also decreased.

There are other mental activities like problem-solving abilities which involve attention and concentration. It is difficult to perceive an attempt at problem solving without having a certain minimal level of attention and concentration. In the L & L Dog IQ Test there are a number of tests which are primarily concerned with problem-solving abilities, but the ability to solve these problems involves a considerable amount of concentration. Examples of these are Test Numbers 4, 6, 9, 11, 13 and 14.

Similarities

Sample item: "In what way are an apple and a banana the same or alike?"

This test measures the subject's capacity for concept formation and for abstract reasoning at the verbal level. When this score is low in comparison with other subtest performances, it often points to some impairment of thought processes or deviance in thinking. The type of conceptual approach (abstract, concrete, etc.) may also point to a certain personality with intellectual traits such as abstruseness, rigidity and lack of contact with reality. The test measures the person's ability to perceive the common elements of the terms he is asked to compare and his ability to bring them under a single concept. A person scoring high on this test may show maturity of thinking. This is seen and recognized in the responses.

A hierarchy of three levels of conceptualization or abstraction may be cited: concrete, functional and abstract or conceptual. Let us assume that we ask a 10-year-old the question, "What is an apple?" If his answer is that it is round, has peelings, a stem and seeds inside, then he has give a concrete response which is the lowest level of abstraction. He wants to put his hands on it and feel it. If we ask another 10-year-old the same question, and he responds, "You eat it," then his abstraction level is higher. He is on a functional or operational level. If we ask a third 10-year-old the same question and he gives us the response, "Fruit," he is operating on the abstraction level. You don't touch or feel the apple nor do you see it when you are functioning on an abstraction level. The concept, fruit, is in our heads.

An example of a similarities question is, "In what way are a dog and a lion the same or alike?" A subject whose thinking is entirely concrete and who is unable to make abstractions is the subject for whom the two have four legs or that the two are not similar. In subjects who are concretistic, though not to this extreme, the response is that the dog and the lion are similar because both have legs or tails or hair. Here concrete detail is taken as the content of the realm although such details are inessential if we deal with concepts of any complexity. In this response a great number of common concrete features of dogs and lions are neglected and not even a complete enumeration of all such details would cover the concept. Besides being too limited, a concrete definition is also too inclusive: tables have legs too. Concrete definitions are therefore inadequate. Another way to define the similarities is to say that both are animals. This is the abstract conceptual level of concept formation. It is the subsuming of items of the realm in question under a general term, the complex content of which is so clearly defined that it is common to our vocabulary and thinking. The response "animals" is a relatively lax one but implies common sense; the response "mammals" may imply sophistication, ostentation, show of information issuing from feeling of insecurity or overmeticulousness issuing from doubt.

Vocabulary

Sample item: "What does 'enormous' mean?"

Vocabulary is word knowledge and therefore reflects not only the subject's capacity to assimilate but also his interest in words. It is thus a measure of the subject's interests as well as of his learning (verbal) ability. While obviously related to schooling and social milieu, the size of one's vocabularly is surprisingly free from the influence of special factors such as age, sex and occupation. It is one of the most stable tests and therefore can be used as the base for appraising intellectual change (deterioration, etc.). The type of definition given is often of diagnostic value, reflecting the subject's type of thinking (abstract, concrete, bizarre, etc.) as well as special preoccupations. This test reflects the richness of the subject's environment and his cultural interests. The size of a man's vocabulary is not only an index of his schooling, but also an excellent measure of his general intelligence. Its excellence as a test of intelligence is seemingly derived from the fact that the number of words a man knows is at once a measure of his learning ability, his fund of verbal information and of the general range of his ideas.

PERFORMANCE SUBTESTS

The performance subtests offer measures of intelligence by evaluating picture arrangements, picture completion, block design, object assembly, digit symbols or coding and mazes.

Picture Arrangement

In this test the subject arranges a series of pictures in the correct order so that there is a meaningful story. The series of pictures are like the short comic strips found in newspapers. The pictures are presented to the subject in a disarranged order and he is asked to put them together in the right order so that they make a sensible story. The test measures social intelligence, planning ability and the ability to comprehend and evaluate a total situation. The subject must get the *idea* of the story before he is able to prepare himself effectively for the task. Correct anticipation is essential for correct performance since it provides for the proper selection of a relevant and appropriate sequence of ideas. The understanding of situations in this test corresponds to what has been generally referred to as *social intelligence*. This test appraises the subject's comprehension and understanding of familiar social situations. Planning and anticipation are two functions often utilized. This planning and anticipation are looked upon as the capacity to look ahead and hence reflect the subject's ability to anticipate consequences of initial action. Unfortunately individuals who are not too well known for their planning ability (psychopaths) also do well in this test.

Picture Completion

This test requires the subject to find and name the missing part of an incompletely drawn picture. He is shown a picture; for example, a watch with its second hand missing and asked to tell what is wrong or missing. Suitable items for this test were hard to find. If one chooses familiar objects, the test becomes too easy. On the other hand, if one selects unfamiliar ones, the test is not a good test of intelligence because special knowledge is required to solve the task. Many pictures were tried out separately and accepted or rejected on the basis of their discriminating value.

The test is good in testing intelligence at the lower levels. It measures the person's perceptual and conceptual abilities in so far as these are involved in the visual recognition and identification of familiar objects. The subject must be able to appreciate that the missing part is in some way essential either to the form or to the function of the object. The factor of sex had to be considered in the selection of items. For example, it was found that more men than women failed to find the missing eyebrow in the picture of a female profile. On the other hand, more women than men failed to find the missing thread in the picture of the electric bulb. Besides serving as a measure of perception, voluntary concentration is sampled. Any disturbances in either the intellectual or emotional areas are reflected in this test. This test measures alertness to environmental details, ability to differentiate essential from non-essential parts in an organized whole and in a general way, visual perception. All are to some extent involved in common everyday reality, and hence reflect the subject's capacity for the same.

Block Design

In this test the subject attempts to reproduce a series of designs shown on a card by putting together blocks. The sides of the blocks are different in color. In the WAIS the test was modified by having all the sides painted red or white, or one-half white and one-half red. This was done partly to eliminate the possible influence of color. Block Design is primarily a test of spatial perception but also calls for abstract capacity so far as this capacity is involved in visual motor organizations. The subject must perceive the pattern, analyze it into its component parts and then construct a given whole. It entails a certain amount of manipulative ability, spatial orientation and a capacity to shift. Wherever these abilities are impaired the subject will tend to do badly on the test. This is especially true of organic cases for the diagnosis of which the Block Design is particularly effective. In other cases it more often reveals the temperamental traits of persistence (the subject's tendency to continue or give up with a difficult design) and degree of frustration. One can learn about the subject's temperamental characteristics by watching *how* he takes to the task. There is also the difference of attitude and emotional reaction on the part of the subject. One can often notice the hasty and impulsive individual from the deliberate and careful type. You can distinguish between a person who gives up easily or becomes disgusted from one who persists and keeps on working even after his time is up.

Object Assembly

In this test the subject puts together jigsaw puzzle items including a manikin, a feature profile, a hand, and an elephant. The elephant, for example, consists of a wide view of a small pachyderm which has been cut up asymmetrically into six parts which the subject is required to piece together. Examiners have said that this test tells them something about the thinking and working habits of the subjects. People with artistic and mechanical skill seem to do well on this test.

This test measures hand-eye coordination. Likewise, it measures the ability to analyze the parts in a given pattern in order to be able to reconstruct the parts. Anticipation of visual-spatial configuration of parts allowing for an organization of these parts into a whole, and the ability to be guided by visual impression in the motor action of the hand in reproducing a pattern are required for good performance. The test has been described as involving perception, motor coordination and practical (manipulative) ability. While it correlates poorly with most of the other tests of the Scale, object assembly stands high as a test of performance. It correlates well with mechanical comprehension. It is useful for evaluating the subject's mode of approach to a set task (trial and error vs. planned attack), as well as reaction to frustration.

Digit Symbol (WAIS) or Coding (WISC)

In this subtest the subject fills in proper code symbols under each number. Even though digit symbol and coding are named differently, they are essentially the same. This type of test measures visual memory, visual-motor coordination in a clerical type task and the ability to learn or adjust to a new situation. The tests are used to measure present learning abilities and motivation. Tests of this kind are not suited for illiterates or individuals with motor-coordination problems.

Mazes (WISC)

In this paper-pencil test the subject is required to trace the right path through a maze. He fails if he enters the wrong path. This test measures planning ability and foresight in the attainment of a goal. Good performance indicates prudent and careful foresight and choice. This test involves problem-solving ability. It is unlikely that a person could solve a difficult problem without both attention and concentration. In some of the more difficult problem-solving items of the L & L Dog IQ Test, it is the opinion of these investigators that concentration is also involved.

Age—Intelligence Scale

The Stanford-Binet Intelligence Scale was a test devised to measure abilities of children at different ages. The method by which an age-intelligence scale is devised is as follows: a group of intellectual tasks of different degrees of difficulty is assembled and administered to children of different age groups. The responses are noted and those test items which tend to discriminate among the subjects are retained and combined into groups usually of six or eight to form various year levels. If there are six tests per year, then each test passed counts two months. Within each group of six or eight forming a year level there are separate aspects of mentality of different mental functions. Wechsler organized his WAIS and WISC in such a way that all the mental functons were in one grouping called subtests, but the items within each grouping were in an increasing order of difficulty. For example, on the vocabulary subtest of the WAIS, the first word the subject is asked to define is *bed* and the last is *travesty*. Sattler (1956) organized the test items of the Stanford-Binet into a classificaiton scheme which permits focusing of the individual test items into meaningful groups. From this perspective we can recognize the mental functions which parallel those of the WAIS and the WISC. Sattler organized seven categories. Each represents a general area conveying the nature of the function measured. The following is a description of each major category.

Language. This category includes items which deal with maturity of vocabulary in regard to the prekindergarten level; extent of vocabulary referring to the number of words the subject can define; quality of vocabulary measured by such tests as abstract words, rhymes, word naming and definitions; and comprehension of verbal relations.

Memory. This category is subclassified into meaningful, non-meaningful and visual memory. Also included are rote auditory memory, ideational memory and attention span.

Conceptual Thinking. This classification is primarily concerned with abstract thinking. Such functions as generalization, assuming and *as if* attitude, conceptual thinking and utilizing a categorical attitude are included.

Reasoning. It is subclassified into verbal and nonverbal reasoning. The verbal absurdity items are the prototype for the category. The pictorial and orientation problems represent a model for the nonverbal reasoning items. Reasoning includes the perception of logical relations, discrimination ability, analysis and synthesis. A spatial reasoning factor is also included in the orientation items.

Numerical Reasoning. This category includes items specifically geared to numerical or arithmetical problems. The content is closely related to school learning. Numerical reasoning includes such factors as concentration and the ability to generalize from numerical data.

The following test items in the Stanford-Binet were taken from year 6 through Superior Adult I levels. They were arbitrarily grouped into the function, Numerical Reasoning:

- VI, 4: Number Concepts
- IX, 5: Making change
- X, 2: Block counting
- XIV, 2: Induction
- XIV, 4: Ingenuity I
- Average Adult 2: Ingenuity I

Fig. 3-1. The act of obeying simple commands does not necessarily reflect social maturity in dog intelligence as it does in human intelligence. (Photo courtesy of Allied Food Inc., Wayne Dog Food.)

- Average Adult 4: Arithmetical Reasoning
- Superior Adult I, 2: Enclosed box problem
- Superior Adult II, 4: Ingenuity I

Visual Motor. This category contains items concerned with manual dexterity, eye-hand coordination and perception of spatial relations. Constructive visual imagery may be involved.

Social Intelligence. This category includes aspects of social maturity and social judgment. The comprehension and finding of reasoning items reflect social judgment, whereas the items concerning obeying simple commands (Fig. 3-1), response to pictures and comparison reflect social maturity.

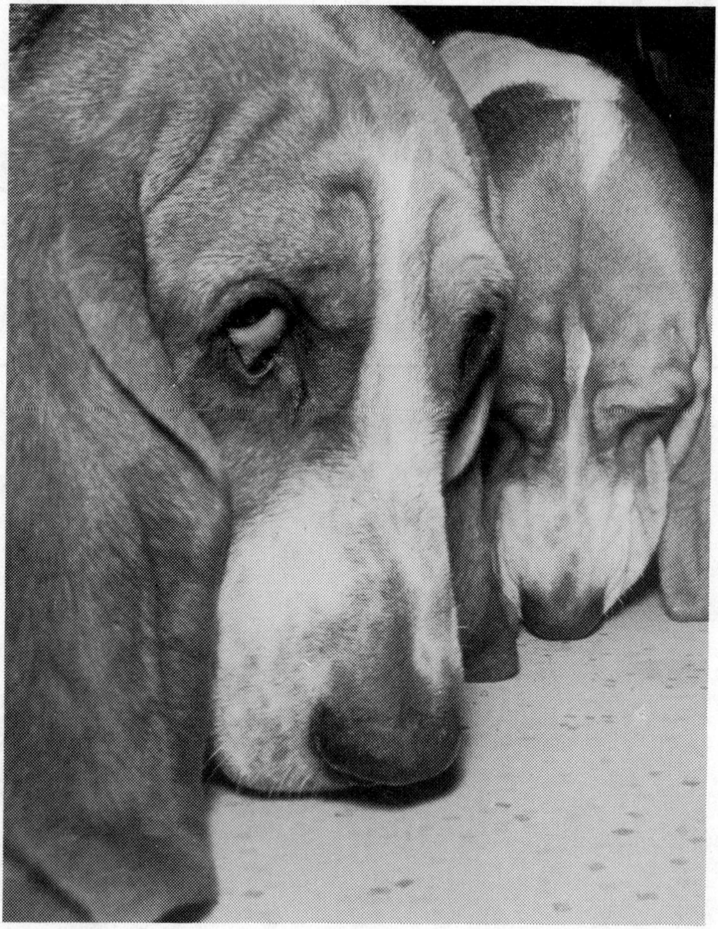

Chapter 4
Animal Intelligence and Its Measurement

What about intelligence as applied to dogs? *Do dogs have intelligence?*

Some researchers in comparative psychology consider that intelligence is nothing more than a learning ability. Others consider intelligence as a reasoning ability. Still others define intelligence as the capacity of an organism to learn to adjust successfully to new and difficult situations. It involves the ability to solve new problems by drawing on past experience. Some psychologists consider reasoning, inference, thinking, insight, problem-solving ability, concept formation and delayed reaction as part of learning.

Those who consider that animals below man are just machines place emphasis on a distinction between humans and other animals in methods of learning. Animals learn primarily by trial and error and conditioning. Unfortunately, man too often uses trial-and-error method to solve problems.

When a dog is confined in a latched box that prevents him from obtaining food outside the box, he attempts to solve the problem. He claws at various parts of the box, tries to push the door open and performs other trial-and-error behavior. After many trial-and-error experiences he finally learns to throw the latch and open the door. Man could solve the same problem very quickly. There appears to be a qualitative difference between men and animals which is labeled as *insightful learning*. It is learning which happens suddenly and as a novel reaction not based on previous experience. Some higher animals seem to be capable of this insightful learning.

Many philosophers and great thinkers have been interested in whether animals like the dog reason and how much intelligence they possess (Fig. 4-1). Aristotle in his *Historia Animalium* held that the dog, the elephant and other higher mammals came close to the mental level of the human child. Descartes, on the other hand, contended that even the highest forms below man are complex living machines with minimum, if any, real intelligence.

LEARNING

Early psychology regarded behavior in terms of instinct. Instinct is valuable only as long as the environment remains constant. But in order to adapt to a changing environment an animal must learn. The ability to survive in a changing environment depends upon the animal's ability to learn. The early psychologists were unwilling to accept the fact that animals can learn.

As mentioned, one definition of intelligence is learning. To understand this definition of intelligence we must understand the meaning of learning. Learning is the possession of knowledge and critical judgment. It can be a highly general term which stands for the relatively enduring change in response to a task-demand. Or, the change can be induced directly by experience. Learning can also be the process by which such change is brought about. Learning is manifested by performance. It is inferred from performance and the conditions antecedent to performance. To learn is to change one's way of acting, to acquire information, to memorize. When you learn a task, you respond to a task-demand or an envionmental pressure in a different way as a result of an earlier response to the same task (practice).

One form of learning is verbal, both receptive and expressive. Receptive language includes the verbal skills that one possesses without having the ability to express the words. For example, item three, at the two year level on the Stanford-Binet Intelligence Scale is *identifying parts of the body*. The examiner shows the paper doll and says, "Show me the dolly's hair." Same for the mouth, feet, ear, nose, hands and eyes. The child is not expected to say the words but indicates that he knows the meaning of the words by pointing. The child is using his receptive language ability. Item Five at the age 2 level of the same test is an expressive language test. It is the *picture vocabulary* consisting of eighteen 2 inch by 4 inch cards with pictures of common objects. The examiner shows the cards one at a time and says, "What's this? What do you call it?" The child is using expressive language when he responds.

Many animals have developed receptive language to a high degree. In teaching a dog to pass obedience training the animal develops receptive language skills. The range of learning ability is different among dogs. Some learn quickly while others take much more time to learn the same task. A more or less uncontrolled factor is the way the dog is taught. If several dogs are taught by the same method, some may learn a procedure like sit-stay in six trials or fewer, whereas another dog may require 50 trials.

An elephant is said to have good learning abilities and develops an excellent fund of receptive language. Zoo keepers, circus trainers and jungle dwellers who use elephants as work animals have long known that the elephant is an intelligent creature. In India people tell fantastic stories about the smartness of the elephant. Rensch (1957) studied the intelligence of elephants in India. He studied the training of elephants. After several weeks of daily training the elephant learns to respond to the instructions, "Stop!," "Go!," "Get up!," "Kneel down!" It takes the elephant several years to learn all the verbal knowledge necessary to work in the forest or in road construction.

One fully trained elephant between 20 and 60 years of age developed an auditory receptive language ability and was able to follow some 20 different commands. The vocabulary included such words as "go forward (datt-datt)," "lift your foot (tol-tol)," the Urdu language of India for "lie down on your side," "push the object with your foot," "push the object with your head" and "squirt water on your back."

Many psychologists consider conditioning to be the fundamental form of learning. Through conditioning the animal's responses to different kinds of stimulus situations are changed. The different ways in which these patterns of responses are acquired has lead to two kinds of conditioning: classical and instrument.

Ivan Pavlov, a Russian physiologist, discovered the conditioned response while doing research on digestion and salivation in dogs. A glass tube emitted the saliva from an opening in the duct of one of a dog's salivary glands. Food was put in the dog's mouth and the salivary response to the food was noted. Pavlov observed that after a number of trials the dog began to salivate when he saw the food. Then the dog would salivate at the sight of the food dish and also at the sound of the assistant's footsteps. Pavlov called the food (meat powder) the unconditioned stimulus (UCS) which brought on the unconditioned response (UCR), the salivation. He showed that after a number of occasions during which a bell was sounded just before the meat powder was put in the dog's mouth, the bell alone produced the flow of saliva. Pavlov named this change in the dog's behavior as a conditioned response (CR) and the bell became the conditioned stimulus (CS).

In classical conditioning the response is initiated by a stimulus (UCS), the meat powder. When learned behavior occurs spontaneously rather than initiated by an external stimulus like the meat powder, psychologists speak of *instrumental conditioning*. This is usually classified as *operant conditioning, escape* or *avoidance*. The food-seeking behavior of the hungry animal is instrumental in obtaining food.

The *Skinner box* is used to study operant conditioning. A hungry animal is put in the box. When the animal presses down on the level, a piece of food drops into the food tray. When the correct response is made (pressing on the lever), the response is immediately rewarded or reinforced. Instrumental conditioning may make use of avoidance or escape. A dog may be placed in a position so that one foot rests on an electrically charged grill. A light is flashed and seconds later the grill is charged with an electric current. This conditions the dog to raise his paw. Later the dog lifts his paw upon seeing the light. There are other forms of learning such as, trial and error, habituation, imprinting, sensory discrimination and motor skill learning.

Another form of learning is termed insightful. Some psychologists equate insight to thinking and reasoning.

Insight learning is the process by which the meaning, significance, pattern or use of an object or situation becomes clear; or the understanding thus gained. In Gestalt theory, insight was originally described as happening suddenly. It was a novel reaction not based on previous experience.

Another definition of insight is: the solution of a problem by a sudden reorganization of behavior. We are dealing with problem solving. An animal is faced with a situation in which it will carry out some act or attain some goal such as obtain food, eliminate an electric shock or obtain a sexual mate. To reach its goal, the animal must perform a series of movements which it has not performed before—even though it has already employed the component movements. An animal may solve a problem by a series of ill-directed movements from which the appropriate ones are gradually selected. Past experience is one factor which reduces randomness. In some cases an animal may solve the problem without delay or error. When the animal does this, it is making use of a previous experience which involves information or the experience of different elements which now make up the whole situation. Its behavior depends in part on the exploratory learning of the different major parts of the apparatus that has taken place on earlier occasions.

Rats display insightful behavior through problems presented in mazes. The following test illustrates trial and error and insightful solution to problems. A chicken or a rat is put behind a wire mesh or glass screen through which it can see food on the other side. To obtain the food it must go around a barrier. A rat seems to fail this test. It will claw at the barrier to get the food and exhibit trial-and-error behavior in an attempt to solve the problem. Finally it begins to run around at random and by chance may get the food. If the animal is placed behind the screen a second time, there is likely to be less trial-and-error behavior and the food will be located in less time. After a number of trials the rat learns to run directly around the screen to the food.

A monkey, on the other hand, will size up the situation quickly. It goes around the screen immediately without trial-and-error behavior. It seems to be able to appraise the situation, put two and two together and solve the problem. The monkey learns by observing first and then acting. Wolgang Kohler (1925) calls this insight behavior.

PROBLEM-SOLVING ABILITY

In the study of animal learning there is a reluctance to grant learning ability to animals low on the evolutionary scale, and to grant problem-solving ability to those higher on the scale. This point of view is personified in Descartes who was willing to grant animals instinctive types of behavior, yet denying them minds. The concept of problem solving started out as a higher mental process. Early experimenters wanted to find out whether animals could do something other than learn. There has been a search for something more than learning, some higher mental processes. Problem-solving is not analogous with these higher mental processes. Decision making is connected with problem solving in that the organism solves a problem when his behavior is blocked in a certain direction. The difference between the two lies in the fact that in decision making the alternatives are available. In problem-solving the alternatives must be created.

Problem-solving ability may be reduced to learning. Truly creative problem solving is made up of former elements of experience, reorganized in such a way to give a new integrated meaning to the whole. A problem situation creates stresses which cause integration of separate experiences. Experiencing difficulty in a problem situation produces direction in the problem-solving behavior which sensitizes the organism to events or objects. These may help in arriving at a solution. A selective process occurs in this direction, depending upon past experience, which picks out the facets of the situation on which the organism concentrates.

A number of techniques have been devised to investigate the problem-solving ability of animals. Thorndike (1898) invented the problem box. It was a cage into which the animal was placed. Somewhere within the cage was a device which if properly manipulated, released a door and allowed the animal to escape. Thorndike observed the slow, gradual improvement in learning. He hardly ever noted any indication of insight. There were seldom any flashes of reason that allowed the animal to solve the problem immediately. He noted that most learning occurred through trial and error. Hobhouse (1901) also worked in this area and devised a number of problems to evaluate the problem-solving ability of animals. One problem was the multiple stick test in which the animal manipulates a short stick nearby to reach a longer stick with which it can obtain a distant food. Another test was the box-climbing test in which a box must be pushed under a food suspended from the ceiling. The animal has to step on the box to reach the food.

REASONING OR INFERENCE

Some psychologists maintain that intelligence is a reasoning ability or inference. The latter is defined as a mental process whereby, on the basis of one or more judgments, a person reaches another judgment regarded as proven or established by the former. An example of reasoning or inference in human beings is Item three, Year XIV of the Standford-Binet Intelligence Scale. The examiner gives a copy of the following problem to the subject and lets him look at it as the examiner reads it aloud. While the examiner is reading, the subject is solving it: "My house was burglarized last Saturday. I was at home all of the morning but out during the afternoon until 5 o'clock. My father left the house at 3 o'clock and my brother was there until 4. At what time did the burglary take place?"

A number of nonverbal techniques have been devised to look into reasoning or inference in animals. One of the simplest of these techniques is the detour problem. In this situation an animal is blocked from the direct approach to food he can see and smell by a barrier or some other procedure. The animal displays reasoning and inference, hence intelligence, when it takes the round about way after a pause filled with nonproductive behavior.

ABILITY TO REMEMBER

After problem solving the next concept which entered into animal learning experimentation was delayed reaction or memory. The delayed response was originally used to test an animal's ability to retain an idea. It is now considered to be one of a number of ways to study memory. In one such test the animal is released in a room with three identical exit doors. Each door has a light near it. The subject learns that if it goes to the door where the electric bulb is lighted, it is rewarded by food found through the passage. If it goes to the unlighted door, it gets a shock from an electric grid. The delayed reaction test is given after the subject learns to go to the right door consistently. The animal sees the light at one of the three doors. The light is turned off and the subject is restrained for a certain period of time before it is released. It is determined how long the animal can remember which light was turned off. A dog will go to the correct door when released provided he is allowed to point his head toward the correct door while waiting. If the dog, however, is turned around it cannot remember the proper location, and performance is seriously impaired. It appears that during the turning around, the dog loses some muscular cues which are important in the correct performance.

Monkeys do extremely well on tests dealing with delayed responses. In one experiment the monkey is shown two inverted cups. While the monkey is looking, a piece of lettuce is placed under one of them. Then the cups are hidden with a screen for a short time or the animal is taken for a walk. After a certain interval the monkey is permitted to go to the cups. The cups are smeared with lettuce juice so that he does not receive odor cues. The animal gets the lettuce only if he selected the correct cup. It is reported that some monkeys can remember the right cup after a delay of 24 hours.

Rensch (1957) tested an elephant's memory. The animal was shown 13 pairs of cars which she had learned earlier but had not seen for approximately a year. In 520 trials she scored between 73 and 100 percent on all the pairs except one. The elephant remembered 24 different visual patterns for about one year's duration. Later other animals were examined for memory. It was learned that the ass and the zebra could not compete with the elephant in the number of stimulus pairs learned. The ass learned only 13 and the zebra 10. The horse, on the other hand, mastered all of the 20 pairs that the elephant had learned. These results suggest that the horse has excellent visual learning memory.

Beritashvili and Aivazashvili (1968) studied short-term memory in dogs after they gained visual perception of the location of food in an experimental chamber. They presented data on the behavior of four dogs. *Short-term memory* was taken as "psychonervous memory, lasting several tens of minutes but not carried over to the next day." When food was shown once or twice a day in any new place, the maximal duration of delay for a correct reaction (running to the place of food) came within 2 hours in most cases. The subject received food many times in the course of the experimental day at various places in the three-experimental chamber. Then after showing the food in only one of these places, the reaction dropped to between 20 and 25 minutes. When food was shown in several places at different times in one experimental situation within the subject's field of vision, then the reaction became considerably shorter. When these places were at a distance of 4 meters, the reaction did not exceed 15 minutes. When at a distance of 1.5 meters, the reaction fell within the range of 5 minutes. They presented a psycho-physiological explanation of the origin of this difference. A correct reaction involved not the "variability of psychonervous perceptual processes or of the image of alimentary location, but the competition of reproduced images of the place of food."

ABILITY TO FORM CONCEPTS

Some research has been done on concept formation and the closely related areas of abstraction and relationship learning in animals. These studies attempt to show that animals respond to a relationship between stimuli. Herrnstein and Loveland (1964) taught pigeons to respond to the presence or absence of humans in photographs. The persons might be anywhere in the photographs. They could be children or adults; black, white or yellow; clothed or nude; or in varied postures. The experiments suggest an abstraction ability in the pigeon.

The concept of triangularlity has been investigated extensively with dogs, rats, cats, raccoons, monkeys and human infants. In one experiment (Fields, 1932) rats were trained to respond to triangles varying in position (apex up, down, to the right, to the left, etc). The only constant aspect of the positive situation was the white equilateral triangle—other concomitants varied. Those animals that learned to discriminate the triangle, regardless of position, were given other kinds of triangles (right-angled, made of lines or dots, etc). They were substituted in varied positions for the triangle used in training. Accuracy was 90 percent or

Fig. 4-1. Many philosophers and great thinkers have been interested in whether animals like the dog reason and how much intelligence they possess. (Photo courtesy of Allied Food Inc., Wayne Dog Food.)

better. The rats reacted to all of these different triangles and positions of triangles as equivalent. They learned *abstraction of triangularity*.

Rensch (1957) working with elephants concluded that these animals could learn abstract concepts. The elephants were first trained to recognize and respond to a black cross. The black cross was turned to the position of an X and the animal responded positively. When the experimenter changed the relative length and width of the arms of the cross, he again responded in the same way as he had to the original cross. Even when a T was made of the cross the animal recognized the figure as a positive sign. The animal got the idea that the essential feature of the pattern was the crossing of two black bars. He did not, however, respond to a white cross on a black background.

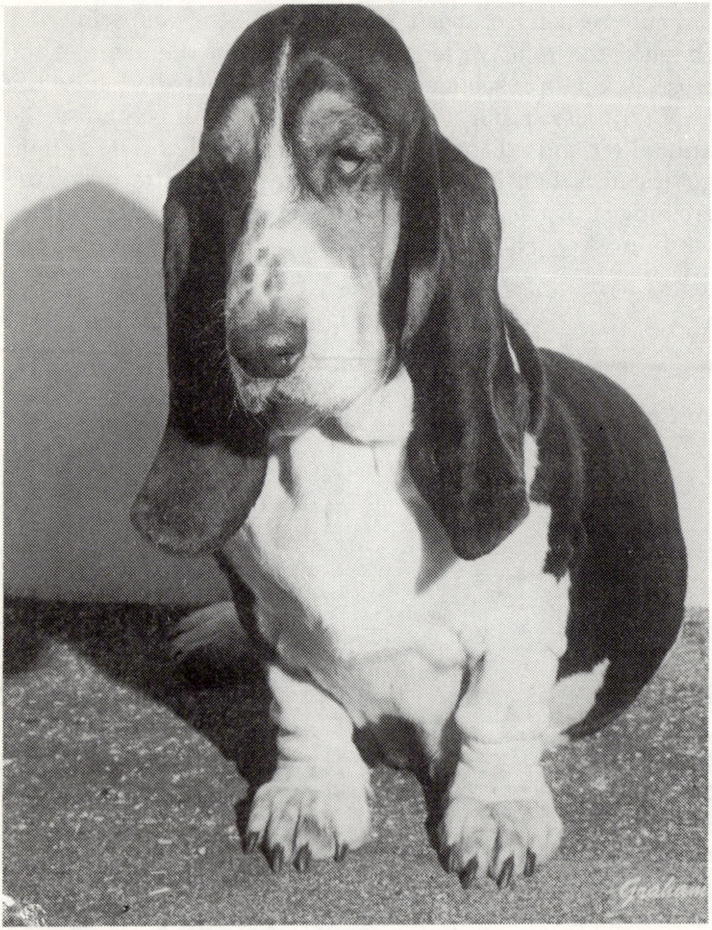

Chapter 5
Intellectual Functions of Dog Intelligence

The intelligence of dogs and the intelligence of humans are widely different, and there are undoubtedly many things about a dog's mental processes that are as yet unknown and others that are poorly understood or misunderstood. There are, however, certain functions which are closely similar to those of humans and which can be understood and measured to a rough degree. As a beginning to understanding dog intelligence, it seems worthwhile to take the applicable functions of human intelligence which have been measured and see how they can be measured in dogs.

FUND OF VERBAL KNOWLEDGE

Two functions that are easily measured in studying human intelligence (WAIS-WISC: Information and Vocabulary) cannot as yet be measured effectively in dogs. Generally, dogs have a good receptive language ability, but it is something that must be developed by their human associates.

The L & L Dog IQ Test contains one receptive language iem. Practically all humans are exposed to language, although this can differ depending on the culture. In dogs, however, there is a much greater variation in the amount of language to which they are exposed. Some dogs seldom hear language except such words as "No!," "Come!," "Good Dog!" and "Bad Dog!" Dogs that are trained and closely associate with people have the opportunity to learn a significant amount of language.

In obtaining training and in training for various kinds of work, dogs can learn many commands. These can be spoken or visual since dogs can be taught to respond to gestures. Here are some of the more common words that dogs learn:

    Heel
    Sit
    Down
    Stand
    Stay
    Retrieve
    Take In
    Fetch
    Bring
    Search
    Find it
    Quiet
    Away
    Wait
    Come
    Go
    Over
    Jump
    Seek
    Hold it
    Out
    Swing
    Back
    Track
    Kennel (to go to)

Each dog also learns his name, and is likely to quickly learn the meaning of "No." On the Cattell Infant Intelligence Scale the infant at 18 months identifies one picture from name. This item is a test of the child's receptive language ability. The child does not need to say or express the word. A card with a picture of a dog, a cup, a shoe and a house is shown to the child and the examiner says, "Where is the doggie? Show me the doggie (bow-wow)." If there is no immediate response the child is asked to "Put your finger on the doggie." This is a test of the child's receptive language ability. Credit is given if the child points to or names one or more pictures. Credit is given at 22 months for pointing out two pictures.

On the Stanford-Binet Intelligence Scale, Form L-M, an expressive language test item is given the child at the 2 year age level. It is known as picture vocabulary. Eighteen 2 inch by 4 inch cards with pictures of common objects are shown one at a time. The examiner says, "What's this?" What do you call it?" The child must succeed on three of the 18 items to pass. Pictures include an airplane, telephone, hat, ball, tree, key, horse, knife, coat, ship, umbrella, foot, flag, cane, arm, pocket knife, pitcher and leaf.

Humans have no dictionary of dog expressive language, but undoubtedly dogs do communicate with each other through sounds and gestures. A person who knows an animal well is often able to get the meaning from such activity. The other day the senior author was walking toward the beach. He noticed that his neighbor's German Shepherd was standing in front of the rear porch door and barking intermittently. The dog's head was raised and he was looking at the door. To him it was saying, "Please, come open the door for me," or "Please come here and feed me," or "Please come and let me in." The owner of this dog would undoubtedly be able to say which of these the dog was "saying."

Our dog, Lore, talks a great deal to us particularly in the early morning. When she thinks it is time for us to get up, she comes beside the bed wagging her tail and thumping it noisily against the side of the bed. At the same time she is opening her mouth and yawning.

There are other kinds of communication between dog and master. Our dog Gertha, for example, has learned to ring the door bell. This was purely a trial-and-error learning. At first she jumped up at the door. Then scratching near the door, she accidentally rang the bell. When the ring was always answered by somebody coming to the door to see who was there, Gertha quickly learned that as soon as she heard the ring, she could stop scratching. She became quite adept at hitting the bell on her first trial.

Receptive language in children and animals is a matter of experience and learning. If a child is totally deaf, he has no receptive language without extensive training. If a dog is confined in a kennel with no human association and is not talked to by his owner and caretaker, he develops no receptive language that is meaningful to humans. Any comparative test of the receptive language of two or more dogs would have to be on dogs that have had similar exposure to people and their language. This is clearly indicated in adult dogs that have been trained in Germany and brought to the United States. When they are exhibited in obedience trails, the handlers generally use the German commands unless they have had the opportunity to put in a great deal of work with the dog to teach him English as well as German. For example, *heel* in German is *zu fuss*. Through training and experience, different dogs have different exposures to language. The words heard by a work dog could be very different from that heard by a pet. Therefore, a different approach to their receptive language seems desirable. The owner should write out a list of words that he thinks his dog knows or might know and test the dog on these. This is illustrated by the samples in Test Number 10.

COMMON SENSE AND JUDGMENT

Dogs seem to have some common sense and judgment (WAIS-WISC: Comprehension) as far as it relates to comfort and survival. For example, many are cautious when close to a bonfire. Some respond quite rapidly to an automobile horn or an approaching car when they are about to cross a street. Many dogs will avoid noxious stimuli. In the L & L Dog IQ Test common sense is determined by presenting the dog with problems which are related to his comfort and survival.

ATTENTION

Dogs have the ability to attend to stimuli. In humans this mental function is measured by asking the subject to repeat numbers given by the examiner (WAIS-WISC: Digit Span). In the L & L Dog IQ Test, a dog's attention is measured using an auditory stimulus (Test Number 7). As mentioned earlier, this test measures how well your brain takes in information through the senses with relatively no effort on your part.

Dog and man possess the same basic senses. The dog must use his basic senses to carry out his role effectively as a working dog. The senses of the dog in order of importance during utilization are smell, hearing, sight and touch. While working, his effectiveness depends primarily upon his sense of smell and hearing. Sight and touch are used least during utilization. However, during training, all of the dog's senses are used.

The primary value of a working dog is his ability to perform as a sensory mechanism. The dog uses his sense of smell to detect odors at great distances. The air a dog breathes contains odorous particles. When this air reaches the portion of the nose associated with the sense of smell, these odorous particles are detected, and sniffing follows. Sniffing enables a generous supply of air to pass into the nose and over the areas richly supplied with the nerves which detect odors. Studies have shown that a dog responds to odor traces of all known sorts and in dilutions far more extreme than can be detected by man. Furthermore, he can distinguish between many odors which seem identical. The dog uses his sense of smell in his role to detect intruders.

The dog's sense of hearing is another reason why he is invaluable when used as a working dog. A dog has a more acute sense of hearing, which enables him to detect sounds that the handler is unable to hear. The dog's sense of hearing is also important because it is the principal medium through which his handler communicates with him. Some dogs appear to understand accurately the feelings and wishes of their handlers as they are conveyed by voice. Usually a word spoken in an encouraging tone, such as "good boy," pleases a dog. A cross word, such as the admonition "NO" tends to depress him. A dog soon learns to associate the sound and tone of a word with the action expected of him.

With one exception—the ability to detect movement—a dog's vision cannot be compared favorably with that of the normal human. To the dog, everything appears to be blurred and out of focus. In addition, he is probably unable to discriminate between colors and sees everything as a black and white or grayish picture. However, he detects an object when it moves slightly and through training, he alerts his handler. Since the dog's vision is limited, it is of least importance. During training, however, his sense of sight is used more when responding to hand gestures.

The dog's sensitivity to touch is primarily used during training when he is being physically praised or corrected. There is a wide variation among dogs in their responsiveness to the sense of touch. A dog's sensitivity to touch can be determined when he is petted or corrected.

CONCENTRATION ABILITY

No animal experiments were found measuring concentration ability per se as reflected in the WAIS-WISC Arithmetic subtest. On this subtest the subject must first understand the problem, retain the problem in his mind for a period of time, determine what arithmetical processes are needed to solve the problem, apply these processes to the problem at hand and come out with the answer. Concentration is the exclusive and persistent attention given to the task being performed. It is a focusing upon the present task and excluding emotions or thoughts unrelated to the task. It is a voluntary effort to solve a problem. In this effort there is an attempt to eliminate interfering internal and external distractions which seek attention and in so doing disrupt concentration. Concentration requires freedom from distractions. It is complete self-involvement in the task at hand. Distraction is any internal or external stimulus that causes a shift in attention and reduces concentration. Humans and animals vary in degree of distractibility. Some distractions are momentary, others may persist for a long time.

In his work with brain-damaged children the senior author became aware of the significance and effects of distractions. One significant characteristic of brain-damaged children is their external distractibility. He would engage them in some interesting psychological tasks. They would start on them, and suddenly they would stop what they are doing and listen to the telephone ring in the next office. He was concentrating on how the child was performing the tasks and was not aware of the ringing until he would attend to the sound. Many of these children also had visual distractions. Frequently while engaged in a task, their attention would wander to some object they saw through the window like an airplane, a car or a person passing by.

Distraction is not a distracton unless it distracts the human or animal. Distractions can be classified as stem-

Fig. 5-1. An apparent concentration by this dog can be observed. (Photo courtesy of Gaines Dog Research Center.)

ming from external, internal or a combination external-internal stimuli. External distractions are those that are brought about by stimuli which result from what is seen,

Fig. 5-2. One of these four dogs is displaying an alert expression of concentration.

heard, touched or smelled. The sight and sound of an airplane above or the sound of a fire siren in the distance may be external distractions to the dog.

Some distractions are completely internal and apparently have no immediate external stimuli to trigger them off. Daydreaming, worrying and irrelevant preoccupations are some internal distractions among humans. There does not appear to be any way to describe and evaluate internal distractions of animals because they don't have expressive language abilities.

Humans appear to have many more ways to handle external distractions than do animals. At times humans use various means to eliminate distractions to which animals might be subjected. For example, persons who have had experience handling horses years ago remember the importance of placing blinders on the animals. The horses pulled carts on streets, and without the blinders they would become startled and distracted by passing traffic. Most cities had ordinances prescribing that horses on city streets would wear blinders to avoid accidents. Perhaps one of the most important ways to decrease the effects of external distractions on animal behavior is to expose the animal to various kinds of distracting conditions so that he becomes accustomed to them.

You can observe concentration in dogs (Fig. 5-1). You have no doubt seen the old Victor logo of a dog with its head cocked to one side listening intently to the music coming out the loudspeaker. You have also probably observed a dog watching something intently as if trying to figure it out. Put a crab on the ground in front of a puppy and the puppy will paw and bat the crab. If the crab pinches the puppy, it will run away yelping. Other puppies will stare at the crab as if trying to figure it out. This staring behavior is probably an indication of concentration, but it is difficult to measure it. Besides, not many people who want to measure their dog's IQ have a crab pond next to their kennel.

My wife enjoys standing in front of the puppy pans and making a noise like a cat or some other sound to which the puppies are unaccustomed. Most of the puppies will cock their heads from one side to the other and display an alert expression suggestive of something more than mere attention and approximating concentration (Fig. 5-2).

In The L & L Dog IQ Test there is no test which measures concentration per se, but Test Numbers 4, 6, 9, 11, 13 and 14 measure primarily problem-solving ability. As part of this mental process it is felt that there is another function measured which resembles concentration.

ABSTRACTION ABILITY

There have been some studies of abstraction (WAIS-WISC: Similarities) ability with dogs and other animals. It seems likely that dogs do have the ability to conceptualize. For example, a dog that has been trained to sniff out a certain kind of substance, like marijuana, would undoubtedly be alert to related substances. In The L & L Dog IQ Test there are no items testing abstraction ability. Such items would require training and would not be practical.

PLANNING ABILITY

The authors could not find any studies dealing with the planning ability (WAIS-WISC: Picture Arrangement) of animals. Some people may think that a dog plans for a future meal when he buries a bone in the same way that squirrels do when they bury nuts. Some dogs have been observed to go back later and dig out the bone, but others appear to forget about the bone entirely. The test battery on dogs' IQs does not include any items dealing with planning ability.

ABILITY TO SEE THINGS AS WRONG OR MISSING

This function (WAIS-WISC: Picture Completion) so far does not seem feasible to measure in a dog, but it is certain that dogs are well aware of some kinds of things which are missing or different. For example, if his bed had been removed or a favorite chair turned upside down, the dog would most likely observe it and call it to the attention of his master in one way or another.

FINE EYE-MUSCLE COORDINATION AND SPACE PERCEPTION

The L & L Dog IQ Test does not include a fine eye-muscle coordination item (WAIS-WISC: Object Assembly and Block Design). Dogs in action generally have excellent coordination and movement in their reactions to space facts.

In jumping from heights, the dog jumps off gracefully and lands on his feet cleverly. Another complex coordination achieved by most dogs is swimming. Some of the space perception can undoubtedly be related to growth, development and experience. In various obedience training work, eye-muscle coordination is highly involved in the jumping tasks.

PROBLEM-SOLVING ABILITY

This test item (WISC: Mazes) is in contrast to a dog problem-solving test which evaluates common sense and judgment. Tests measuring problem-solving ability in hu-

Fig. 5-3. Change in behavior is often noticed when a dog is exposed to a new environment, such as a lake or a pond. (Photo courtesy of Allied Food Inc., Wayne Dog Food.)

mans do not contain threats to survival or marked discomfort as they do in dogs. Test Numbers 2, 8, 12 and 16 are problem-solving items included in the battery which contain minimal concentration.

ABILITY TO ADJUST TO A NEW SITUATION AND MOTIVATION

Motivation (WAIS: Digit Symbol; WISC: Coding) can be evaluated subjectively by a dog's eagerness of response. For example, on the recall, which is part of an obedience trail, a dog is supposed to respond quickly and eagerly.Some dogs, however, will walk at a slow gate and act almost as if they do not care. A tiny dog will often run for cover to escape a huge dog. The small dog is adjusting to a new situation quickly. Changed behavior of this kind is an example of adjusting to a new environment (Fig. 5-3). The authors have not yet been able to develop a practical test to measure this function.

Questions have arisen in the past regarding the validity of a test which has been abbreviated by eliminating items or certain subtests. The undesirability of such an abbreviated form was clearly established through research by Luszki, Schultz, Laywell and Daws (1970). However, this test for dogs is a somewhat different situation because only the general pattern of the WAIS-WISC is used. Some of the WAIS-WISC subtests are totally inappropriate for use with dogs. Other subtests require very substantial modification.

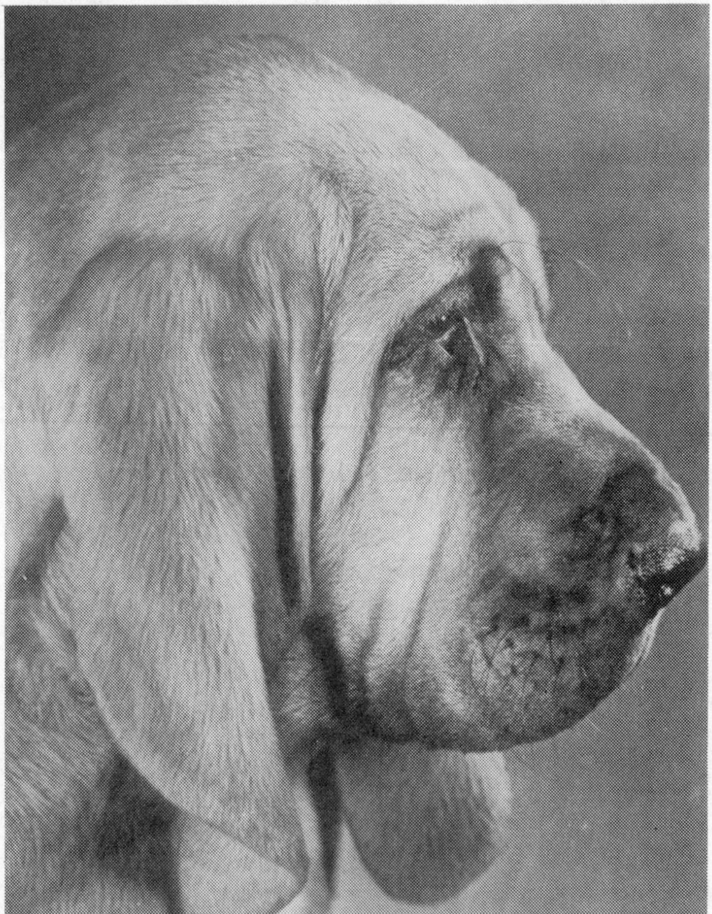

Chapter 6
Effects of Deprivation and Enrichment

Can you make your dog smarter? Can you increase his IQ? Or can you expect a dog to have about the same IQ year after year? In the case of human intelligence, the general conclusion of a large number of researchers is that the IQ remains essentially constant when the environmental conditions are about the same. This includes health, type of education and the home situation. This does not apply to very young children whose potentialities are much more subject to change as a result of different environment influences. Where there is an unusual change in the environment, there can be an increase in the IQ.

HUMAN INFANTS VS. PUPPIES

A number of experiments have focused on environmental deprivation versus enriched environment. These studies have implications for understanding and improving dog intelligence. The average IQ of children such as the Kentucky mountaineers, who are isolated from normal education opportunities, is below the average for children who receive normal schooling (Wheeler). The IQ of these children usually declines with age. When these children are given normal educational opportunities, they often show a marked increase in IQ despite their initial handicap. It is very likely that a similar condition exists among dogs. A dog reared in a deprived environment has little opportunity to develop skills. A dog with the same basic genetic endowment reared in an enriched environment and trained in various skills will be much smarter (Figs. 6-1 and 6-2).

Children who attend nursery school for a year or two show an increase in IQ. Some studies have shown an average increase of seven points after two years of nursery school attendence. It appears that nursery school children have educational opportunities not enjoyed by children on whom IQ tests were standardized. What is the basis for this rise in IQ points? A child who has been in nursery school for a year or more usually has learned many important skills. The nursery school child learns earlier than the non-nursery student according to Wellman. The nursery student goes through an enrichment program. He learns how to cooperate with adults other than his parents. This is an opportunity missed by most non-nursery school children.

Many nursery school students are given tests of different kinds and this experience may make them accustomed to testing situations in later life. Often researchers use nursery school students as subjects in experiments, and this gives the nursery school students additional contacts. Nursery schools invariably provide enriching experiences in such activities as building with blocks, listening to stories being read, listening to recordings, threading beads, putting together jigsaw and other puzzles, learning various group games, learning how to associate with teachers and peers and many other things.

Just as children who are exposed to an enriched environment have an opportunity to raise their IQ, dogs too can increase their intelligence through a program of enrichment, involving exposure to a variety of experiences and training (Fig. 6-3 and 6-4).

D.G. and J.W. Forgays studied the nature of the effect of free-environment experience in the rat. The adult problem-solving performance of rats raised in a free environment with and without playthings were compared with those animals which were raised in restricted and normal environments. The results favored the free-environment group with a high degree of statistical significance. The differences were interpreted in this way due to the differential opportunity for visual learning during early life. All rats were

Fig. 6-1. Having been reared in an enriched environment, this Standard Poodle quickly learns various skills. (Photo courtesy of Gaines Dog Research Center.)

hooded and placed in a four-story cage (5 feet by 2 feet by 2 feet) under four conditions:

- free-environmental rats; playthings available; two mess cages with three rats in each

Fig. 6-2. This Poodle is easily trained in various skills. (Photo courtesy of Gaines Dog Research Center.)

Fig. 6-3. This Golden Retriever's intelligence has been increased by a program of enrichment. (Photo courtesy of Gaines Dog Research Center.)

Fig. 6-4. The program of enrichment for this Golden Retriever includes an exposure to a variety of experiences and training. (Photo courtesy of Gaines Dog Research Center.)

Fig. 6-5. This Pointer has the opportunity of exploring different smells. (Photo courtesy of Gaines Dog Research Center.)

- free-environmental rats; two mess cages with three rats in each
- playthings; two mess cages with three rats in each
- two mess cages with three rats in each

Another important reason for early training in nursery school programs is that habits form early in life are especially persistent in adulthood. There are several reasons for this which are pertinent to human learning. Certain physiological needs and their corresponding drives may be more intense in infancy than in adulthood. Because of this greater intensity under which the experience originally occurred, the effects of early experience might be expected to persist, while the effects of later experience may not persist. Therefore, if the infant does not have opportunity to learn certain kinds of information, attitudes and behavior at least at the time they are normally acquired, he may have difficulty in learning them in later life. The effects of experiences in infancy are consistent with the behavior of animals and have contributed to an understanding of the probable consequences of deprivation experiences.

Because of the rapid development of dogs it is particularly important to start socialization and training at an early age. Begin playing with them and introduce them to new experiences and stimuli—balls, other toys, mirror, whistles, household noises, different kinds of people, stairs, doors, gates, etc. Put the dog in situations where he has an opportunity to explore different smells (Figs. 6-5 and 6-6). He also needs the opportunity to investigate the source of sounds (Fig. 6-7). Give the dog the experience of walking on different footing. Introduce him to a pond (Fig. 6-8), stream, or even the ocean, if this is possible. Have the dog ride in a car (Fig. 6-9). Introduce the dog to different children (Fig. 6-10) and adults (Fig. 6-11) so that he will learn to stand his ground calmly and not cower from strangers.

A review of the literature on human development indicates that the aspect of early deprivation which has the most deleterious effect on later development and functioning is deprivation which interferes with the relationship between the young and its mother. Studies have focused on a variety of different effects on perceptual, cognitive, motivational, emotional and other factors.

Maternal deprivation or lack of mothering involving decreased human contact results primarily in decreased tactile, kinesthetic, auditory, visual and emotional stimulation (contact comfort, warmth, tenderness and affection). Most of the literature dealing with mother-child relation-

Fig. 6-6. Exploring different smells is vital to the training of this German Shorthair Pointer. (Photo courtesy of Allied Food Inc., Wayne Dog Food.)

Fig. 6-7. Investigating the source of sounds will help increase the intelligence of this Great Dane. (Photo courtesy of M.P. Weick.)

ships stresses the importance of adequate mothering and the ill effects of the lack of it. This view holds that institutional infants have higher mortality and morbidity rates than home-reared infants. They also have an inferior development of intellectual, emotional and physical aspects of personality. In contrast to this view some investigators contend that infants brought up in institutions develop normally if the type of adult caring is adequate, and they are provided with sufficient stimulating materials.

Margaret Ribble noted that deprivation of mothering seemed to be related to negativism, depressive conditions,

Fig. 6-8. This Greyhound is introduced to a pond. (Photo courtesy of Carole Kessler.)

regressive tendencies, stupurous states and retardation of the central nervous system, particularly the sensory-motor integration of vision, hearing and grasping. Mothering embraces three factors:

- continuance of the closeness of the prenatal state
- Small acts by which the mother consistently shows her love for the child (these include not only the details of physical care, but also evidence of tender feeling and affection)
- understanding and meeting the infant's biological needs.

Fig. 6-9. Riding in a car offers a new experience for this Bulldog. (Photo courtesy of Gaines Dog Research Center.)

Fig. 6-10. Introducing this Beagle to children will help him learn to stand his ground calmly and not cower from strangers. (Photo courtesy of Allied Food Inc., Wayne Dog Food.)

Fig. 6-11. These German Shepherds are being acquainted with an adult to prevent them from cowering from strangers.

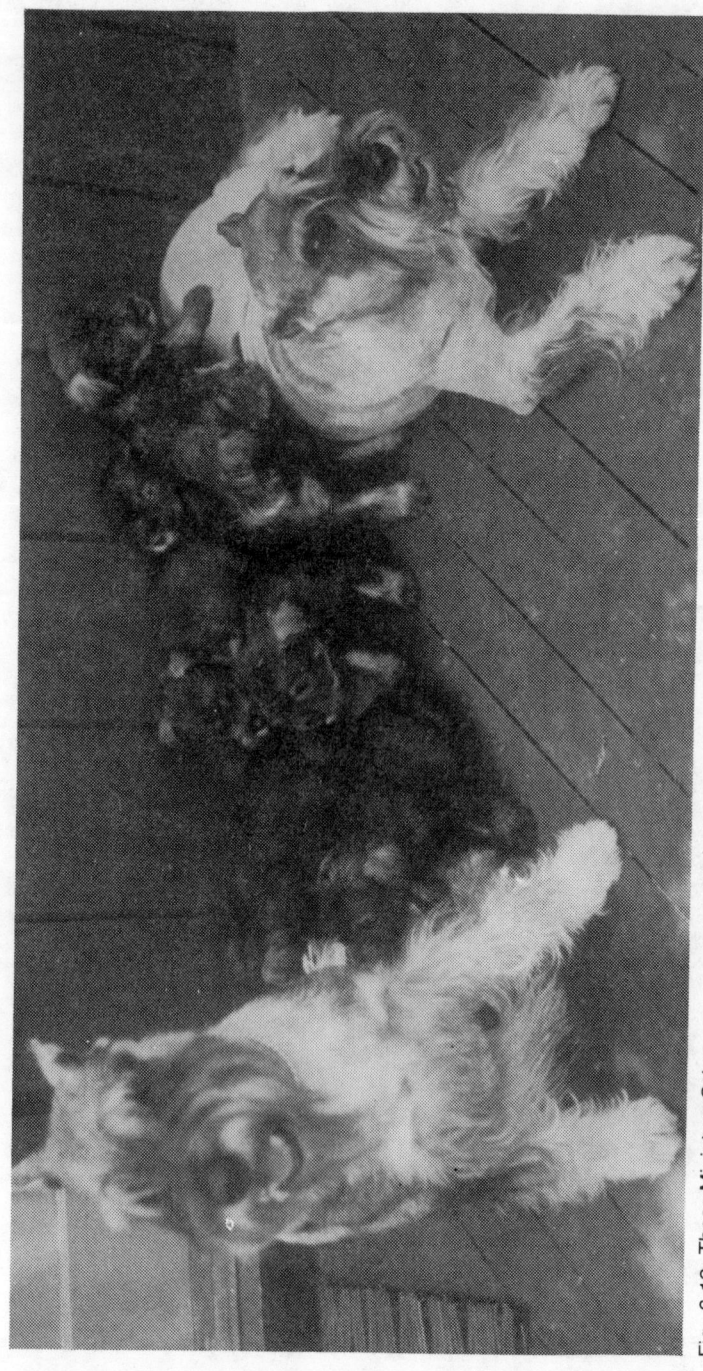

Fig. 6-12. These Miniature Schnauzer puppies exhibit their sucking reflex. (Photo courtesy of Gaines Dog Research Center.)

The biological needs are nutrition, respiration and sensitivity. Food is an obvious need. Hunger for oxygen is another. The breathing mechanism is not complete at birth and respiration remains a problem for many weeks. The brain cells need an enormous amount of oxygen at this time. An oxygen shortage may seriously and permanently damage the brain. The process of breathing is aided by handling and fondling the infant, since his first reflex to touch is respiratory in nature. The child also undergoes respiratory crying. Good breathing insures a supply of oxygen and also determines smooth speech development. Throughout life this is related closely with both physical and mental health. Related to respiration and nutritional needs is the need for sucking. Sucking becomes associated with other senses with which the infant sees, hears and touches the mother while nursing.

Levy studied the sucking reflex (Fig. 6-12) and social behavior of dogs. Two dogs drank from small-holed nipples and sucked on the average for 80 minutes per day. Two other dogs sucked on large-holed nipples and sucked on the average for only 13 minutes daily. He found that the dogs sucking for the shorter time periods were in a state of tension. They excessively sucked and licked immediately after sucking. Also they were more restless, more active and slept less. In addition, these dogs had their jaws affected due to lack of exercise of jaw muscles. They were also poor in general activity.

Fig. 6-13. Three-week-old Scottish Terrier puppies still view their mother as the only essential living object. (Photo courtesy of Mary Chisolm.)

A fairly common practice in recent years for puppies whose mothers are unable to nurse them is tube feeding. A tube is slipped into the puppy's mouth and down into the stomach, and a substitute for mother's milk is given. This works well nutritionally, but when it is used as a complete substitute for breast or bottle feeding, the puppy loses out on the opportunity for sucking. It seems quite possible that it would have adverse reactions similar to those of the human infant. Tube feeding would appear satisfactory in the case of a bitch who has more puppies than she can feed adequately as long as each puppy in the large litter has some opportunity to nurse.

Another major need of the human infant is in the realm of tactile sensitivity. To bring out the sensitivity of the skin, stimulation of the entire skin surface by baths and rubs must be provided. The infant's kinesthetic sense can be developed by rocking and handling. In the case of puppies this need is cared for by the mother who licks the puppies, rolls them over and pushes them around the welping box.

Another basic human need is that of sleep. These needs are interdependent and must be met by the mother if the child is to develop adequately. She must stimulate and direct the activities of the baby. Thwarting of any of these needs disrupts the entire system of the infant. The mother, according to Ribble, must be consistent so that the baby expects and depends on this love. To maintain a feeling of security all needs must be met until the time when the child can communicate by means other than crying.

R. Spitz also showed that there was a deterioration and lack of development of children due to the lack of mothering. The investigator compared children reared in foundling homes to those reared in a nursery situation. In the foundling home environment there was only one nurse for every eight infants. Each child was in a separate cubicle and there was no vision of other children. In the nursery situation there was one mother or substitute for each child. Each child in this cubicle could see others. At 6 months each child could go into the wards with five or six others in the room. Toys were available and there was some competition among the mothers about the babies.

Although the foundling home children came from superior stock, Spitz feels the their inferior development was related to the lack of the mother-child relationship. The mother or other human partner is necessary for emotion.

With puppies the mother is *the* essential living object up to four weeks (Fig. 6-13). At this stage they are able to move around independently. The mother is still important, but they take more interest in their litter mates as well as other animals and people with whom they come in contact (Fig. 6-14). If taken away from the mother before about 7 weeks, there is likely to be a lower level of development in adulthood than if the puppy spent the first 7 weeks with the mother and other dogs.

R.J. Levy studied infants awaiting adoption. Her main sample was composed of 122 babies. Eighty-three were cared for in an institution and 39 were cared for in foster homes. All of them had come into the agency's care within their first 2 months of life. They had been tested around 6 months of age. Those in the institution were in one large nursery, which had accommodated 17 babies and was staffed by a total of 10 practical nurses. There were never fewer than two nurses in attendance during the day. The foster home children were significantly higher in various abilities, including intelligence, than the institutionalized children.

One of the most recent studies of early-environment relations is that of Drs. Marshall H. Klaus and John H.

Fig. 6-14. Already, a couple of these Miniature Schnauzer puppies are beginning to take an interest in things other than their mother. (Photo courtesy of Gaines Dog Research Center.)

Kennell of Case Western Reserve Medical Center in Cleveland. They showed that there is a *sensitive period* during the first minutes and hours of life in which the infant has maximum opportunity for later development if the mother has close contact with her infant. At 1, 3, 6, 9 and 12 months the baby who spent hours immediately after birth with his mother gained more weight and had fewer infections. The mothers were much more affectionate. When the long-contact children were tested at 2 and again at 5 years of age, they showed higher IQs and superior scores on language tests than children who had only brief contacts with their mother at birth.

Many dog breeders have made related observations, namely, that if the mother rejects the puppies that have been taken from her prematurely, either temporarily or permanently, these puppies have a much poorer start in life. They probably have lower intellectual development, although to the knowledge of these authors this aspect has not been studied. Numerous other studies including those of H. Rheingold, W. Goldfarb, L. Kanner and John Bowlby confirm these findings.

Somewhat different from the above are the studies dealing with imprinting. These also shed light on the mother-child relationship and affects of early associatons. E.H. Hess conducted a number of experiments on imprinting, a term coined by K. Lorenz who described the phenomenon as the establishment of a releaser stimulus during an early critical period. Lorenz discovered that ducklings who were in contact with a human during an early critical period would continue to follow that human from then on and would have nothing to do with other ducks or with their own mother. Imprinting has been observed in insects, fish, sheep, deer and buffalo.

Hess studied ducklings. They were kept away from all stimuli since the time of hatching in isolated incubators. A duck decoy, equipped with a loudspeaker and a heating element, is rotated in a 5½-foot circular runway. The loudspeaker can emit a human rendition of *gock*. Ducklings were placed in the runway at different time periods following birth and permitted to follow the decoy. The critical period for maximum response was 13 to 16 hours (range: from 5 to 24 hours). Under natural conditions imprinted ducklings avoided their mother and moved closer to the decoy, while naive ducklings over a day old avoided the decoy and went to the real mother.

The imprinting concept suggests that infants born in hospitals are doomed to experience initial frustration, since they really prefer the nurse who cared for them but are forced to tolerate their parents. The imprinting concept also points to the importance of mother-raised puppies rather than puppies that are rasied by a substitute non-canine mother.

Some experiments have been conducted emphasizing factors other than social relations (Fig. 6-15). Various studies suggest that in some species of lower mammals, birds and primates, prolonged visual deprivation during the early developmental period may result in difficulty in reacting adaptively to visual cues when the animal is removed from the deprivation experience.

TESTING OF OTHER ANIMALS

A.H. Riesen and his co-workers found that marked increase of stimulation among organisms reared in a restricted environment lead to certain perceptual deficits and emotional storms. Working with cats, monkeys and chimpanzees, he concluded that excessive arousal effects of stimulation after early sensory deprivation resulted in perceptual deficiencies, hyperexcitability, increased suscepti-

Fig. 6-15. Basset Hounds enjoy social relations with another breed. (Photo courtesy of Gaines Dog Research Center.)

bility to conculsive disorders and localized motor dysfunctions. He found that by minimizing visual and tactual-kinesthetic contact with the external spatial environment, kittens and apes at 4 to 12 months became emotionally excited in a degree according to the constellation of stimuli given. The change in environment led to fearful withdrawal or emotional storm. A cat was born in a dark cage and kept there for 1 year. From 1 to 2½ years it was brought into a variegated and well-lighted environment. The visual motor environment was meaningless for a long time. The cat trembled, meowed or growled in fear of the strange situation. The cat ignored other cats and withdrew from them if they became aggressive. At the age of 3 the cat was still unadaptive to spatial detour problems. Riesen believes that individual, social and physical environmental factors all interact to determine how many early deficiencies the cat can overcome when given such opportunities in the second and third year of life. There would probably be similar findings if puppies had been used in a study of this kind.

D.O. Hebb used rats in studying the effects of early experience on learning and problem solving at maturity. He used two methods. He compared animals blinded at birth with animals blinded at maturity, and he limited the environment by comparing animals reared to maturity in cages with those which were given varied experiences. In the latter method he raised one group of rats as pets in a private home, while another was raised in the laboratory. The former did better on the T-maze. He concluded that there is a lasting effect of early life experiences on the problem-solving ability of the adult rat. Animals that are treated like pets do better at perceptual tasks than animals raised in isolation. This would undoubtedly apply to dogs too.

Bernard Hymovitch made a follow-up study on Hebb's experiment. He conducted three experiments on the effect of altering early experience patterns on later performance and learning. In experiment I the rates were blinded in early life or at maturity and reared either in free environment or in normal cages. Results: on a closed field test at maturity it was indicated that the free-environment rats were superior to the normal cage rats. Differences between early and late blinded groups were not significant. In experiment II rats were reared in individual stovepipe cages, in mesh cages, in enclosed activity wheels (gives muscular exercise) or in a free-environment box. The mesh cage and free-environment groups were superior to the stovepipe and activity wheel group, but differences between mesh cage and free-environment groups and between stovepipe and activity wheel groups were not significant. In experiment III one group was given the free-environment experience during early life and restricted to the stovepipe cages later and another group was restricted to stovepipe cages in early life and given free-environment experiences later. The first group was conclusively superior. Hymovitch explains the differences of wider, early perceptual learning, although early versus late blinding had little effect. The enclosed maze scores were not affected by the early experiences.

R.S. Clarke and his colleagues raised three Scotties with practically no human contact and in a cage out of which they could not see. At the age of 3 months they were removed and placed with three Scotties who had been raised normally. The dogs raised in isolation disliked being handled. They froze and hugged the floor. They also didn't complain when given injections. The other dogs were normal. Six months later the gross symptoms had disappeared, but there were still evidences of reactions similar to those found by Spitz.

Some research has been on environmental deprivation regarding the availability or nonavalability of materials that can be manipulated. F.A. Beach showed that female rats, although they had never seen nests made were nonetheless able to make use of nesting material. This nest building behavior seems to be related to earlier experiences in handling things within the environment. The female rats that had never delivered litters before were also able to clean and care for their young. B.F. Riess found that female rats reared in cages which did not contain anything they could manipulate did not build nests when the young were born even though nesting material was available.

Some experimentation has been done on deprivation of perceptual experiences. Many experiments have borne out Hebb's theory that animals that had a large amount of perceptual experience early in life will be better learners than those deprived of such experiences. Also, the earlier the animals gain this perceptual experience the greater will be the facilitative effect on later learning.

Fig. 6-16. The environment affects the intelligence of dogs. (Photo courtesy of Gaines Dog Research Center.)

W.E. Bingham and W.J. Griffiths, Jr. used rats as subjects. From age 21 days, experimental rats were reared in a specially designed room and given access over a period of 30 days to tunnels, inclined planes and swinging doors. Controlled animals were reared in restricted *squeeze boxes*. At adulthood the two groups were compared by using mazes, jumping apparatus and open-field temperament tests. Susceptibility in sound-induced convulsions were also studied. It was concluded that rats reared in wider and richer environmental rooms were superior in maze learning ability to animals reared in the squeeze boxes. There were no differences traceable to early experiences in temperament, discriminating behavior or susceptibility to sound-induced seizures during adulthood.

The general outcome of studies on perception seems to be that the fundamental perceptual skills must be learned early in life. Serious deprivation in early life might result in impairment in these areas.

All of these studies, whether they relate directly to dogs, to humans or to other animals, point out the importance of early environmental influences (Fig. 6-16) and experiences. These factors help raise the IQ and improve other aspects of development.

One fairly common situation found with dogs is overpampering by owners. The dog becomes a "spoiled brat". He takes over the household, does what he wants to do and like the demanding, spoiled child, follows his own whims rather

Fig. 6-17. It is important not to spoil your dog by letting him take over your household. (Photo courtesy of Mrs. Katherine Layton.)

than any rules that the too permissive owners try to establish. A dog who has been spoiled (Fig. 6-17) in this way is very difficult to train because he wants to do what he wants to do. He never achieves his intellectual potential.

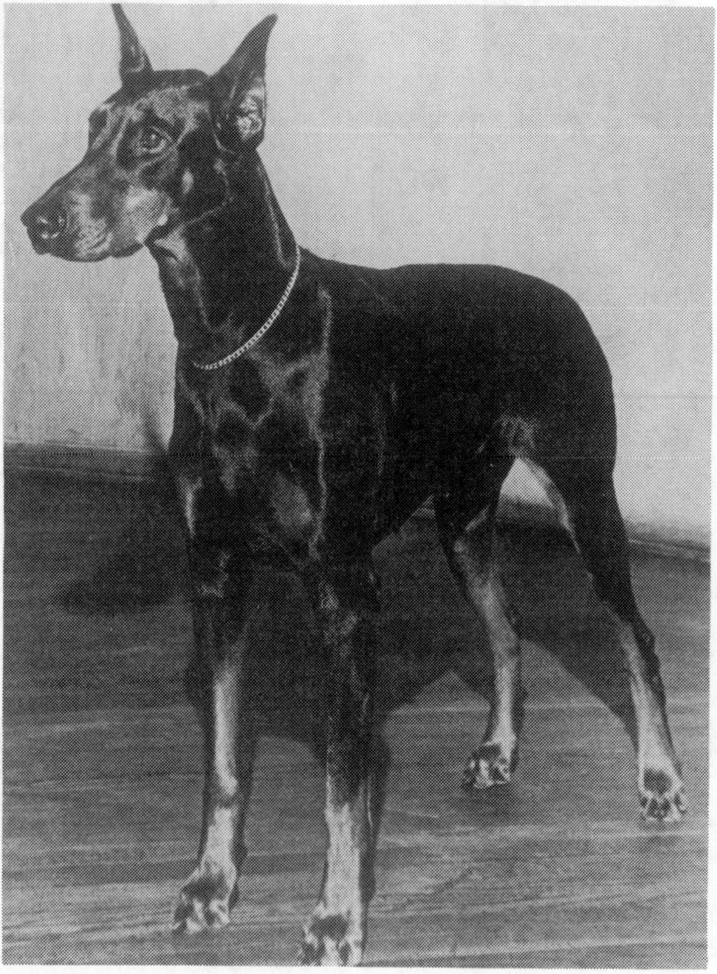

Chapter 7
Sources of Human and Dog Intelligence

For almost more years than we can remember there has been controversy about the relative role of heredity and environment in human intelligence. Both are important. They are equally important in canine intelligence. Let's turn away for a minute from intelligence as measured by intelligence tests, such as we will be talking about in the next two chapters, and look at it more generally. We like one of Webster's definitions: "The power of meeting any situation, especially a novel situation, successfully by proper behavior adjustments; also, the ability to apprehend the interrelationships of presented facts in such a way as to guide action towards a desired goal."

The intelligence tests we will be talking about try to do this, but the definition suggests the difficulty of any accurate and standardized measure of intelligence which will be accurate and fair to all people—or to all dogs in every breed.

EFFECTS OF THE ENVIRONMENT

Intelligence is influenced by both heredity and environment. Many research studies on humans have been conducted to find the determinants of intelligence and to assess the relative importance of each of these variables. We can study the effect of environment on intelligence by placing a person in a different environment but holding the hereditary factor constant.

One way to hold the hereditary factor constant is to study a number of identical twins. Such a test was conducted and each twin was placed in a different foster home. After several years psychological tests were given to the subjects and the scores of one twin compared with the scores of the other. The heredity of identical twins is the same, and if there are differences between the twins it can be assumed that the differences can be accounted for by environmental factors. The typical finding in twin studies is the IQ changes to some extent with changes in environment, especially at an early age. However, the IQ tends to remain constant, unless there is a dramatic or drastic change in the person's environment.

An example of a dramatic kind of change is that of a 2-year-old child living with a mentally retarded mother in a primitive environment. If this child is adopted by two highly intelligent persons in an enriched urban environment, this change may increase the child's IQ by many points over a period of several years. The marked change in the environment may account for the difference in intellectual development.

Other studies compared the performance of identical twins, who have the same heredity, with that of fraternal twins, whose heredity is no more similar than that of ordinary sisters and brothers. One such study by Blewett (1954) showed that the correlations between the scores of identical twins were much higher than between those of fraternal twins on the Thurstone's Chicago Tests as a whole. They were also higher on each of the primary mental abilities except numbers, on which the identical twins made only slightly more similar scores than the fraternal twins, indicating that numerical ability depends considerably on environment. The highest correlations were obtained between the identical twins on the verbal and fluency factors, indicating that these may be highly hereditary.

Deprivation

Early deprivation studies of human infants confirm the conclusions drawn from animal investigations. Spitz (1945, 1946), Spitz and Wolfe (1946), Kanner (1949) and others have shown the relationship between early life experiences and intellectual development. Long before words are meaningful to an infant, sounds stimulate him. The deaf infant lacks this stimulation. Not only is he deprived of the opportunity to learn verbal symbolism, but his perceptual processess and other nonverbal behavior are established and structured differently from those of the child with normal hearing. When the deprivation of both verbal and nonverbal auditory experiences of the deaf child are considered, there is little doubt that, unless strong and soundly based remedial action is taken, the deaf child's intellectual development will be less than it would have been if the child had not been born deaf. Both the limited inputs he receives and the lack of feedback to permit the correction of errors contribute to slower and more limited intellectual development.

Concept development is particularly likely to be impaired in the deaf child. Long before words become meaningful, there are associations between sound and concepts. Some sounds remind the baby of his rattle and others tell him he is with somebody. Sounds stimulate him intellectually, socially and emotionally.

Myklebust (1960) has set up hierarchies of experience ranging from concreteness to abstraction, categorized into the levels of sensation, perception, imagery, symbolization and conceptualization. If the level of hearing sensation is impaired, then all categories above that level will be altered. The deaf child is likely to have different perceptions, imagery, symbols and concepts from the normal child.

Restrictive Surroundings

Environmental influences at the beginning of life shape the behavior of animals other than humans. Thompson and Melzack (1956) attempted to learn how an animal is affected by severely restricting its chances for development and learning during its first few months of life. These researchers wanted to learn whether restricted surroundings during the early rearing of Scottish Terriers resulted in deleterious effects upon the dog's intelligence, activity, emotional reactions and social behavior. The research was conducted for five years in the psychological laboratory of McGill University. At the age of four weeks when the Scotties were weaned, each litter was divided into groups: controls and experimentals. The controls went to Montreal homes or as pups in the laboratory. The experimentals were confined to cages. Each of the experimental dogs was put in a cage which was closed in with opaque sides and in some instances an opaque top. The dog could not see outside. Feeding and cleaning were accomplished in such a way that the dog did not see its keepers. The dogs were in this deprived environment until they were between seven and 10 months old. Then they were let out of their cages and given the same handling and exercise as the controls. The two groups were then given a number of psychological tests.

When the experimental dogs were released, they were active and playful rather than showing a depressed affect. They were placed in a small room and observed for 30 minutes to see how much time they spent exploring the room compared to sitting or lying down. The normal puppies soon became bored and relaxed, but the experimentals went on exploring for a considerably longer period of time. Also, older dogs became bored quicker than younger ones.

The experimental and the normal dogs were invited to explore a maze. Initially both groups were equally curious, but the experimentals ran about more actively than the controls. This behavior persisted several years after they had left their early cages. This behavior was interpreted to mean that a rich early life decreases over-curiosity. The rich early life produces an animal which is just as curious about a new situation, but it has the means to satisfy its curiosity quickly. It was determined that the experienced animal showed more intelligence.

In a learning experiment involving pain-avoidance, the normal dogs did better than the restricted dogs. In this experiment the dogs were chased by a toy car which gave them a mild electric shock. The car was controlled remotely by the experimenter. The normal dog quickly learned to avoid the car without behaving widely and aimlessly. The restricted dog took much longer to learn to avoid the car and showed more excitement.

A number of other tests of intelligence were conducted to evaluate the differences between the normal and the restricted dogs. In one test the dogs were trained to get food by running along a wall of a room from one corner to the next. The food was then placed in some other corner of the room. The experimenter made sure that the dog saw this change in food location by banging the food pan on the floor to attract the dog's attention. The normal dogs usually went to the new corner, but the restricted dogs almost always ran toward the old corner.

In another test of intelligence a chicken-wire barrier was placed in front of the food to see whether the dog would go around the barrier. The restricted dogs showed unintelligent behavior. They dashed to the barrier, pawed at it and tried to push their muzzles through the wire time after time. The normal dogs learned to go around the barrier after one or two trials.

A delayed response test was given to measure the dogs' skill in observing and remembering the location of food. Each dog watched the experimenter put some food in one of a pair of boxes and was permitted to smell the food. Then the dog was taken away and after a short time released to test whether he could remember the correct box. The normal dogs did much better than the restricted ones. Generally, the restricted dogs could not pick the correct box reliably even when released immediately.

A set of 18 different maze problems were included in the intellectual evaluation. The dogs were first trained to do simple mazes. The conditions were somewhat similar to those used for human intelligence evaluations. The subjects taking the tests had previous training in running mazes similar to those used in the test. The dogs were scored on the number of errors and how long it took them to solve each maze by reaching the food at the end of the run. The normal dogs did much better than the restricted ones. The restricted Scotties made about 50 percent more errors than the normals.

INHERITED CHARACTERISTICS

It is as true in dogs as it is in people that intelligence is in part a result of heredity. Inherited characteristics are induced by the actions of genes—tiny units in the chromosomes of the body cells. Every body tissue is made up of cells. Each cell contains a nucleus and the substance around it, the cytoplasm. Within each nucleus are the chromosomes, and within each chromosome are the genes. They transmit the inheritance. In dogs, the normal number of chromosomes in body cells is 78 (in man there are 46). As far as we know, there has been no determination of the total number of genes in dogs. In the spermatazoa and the ova, the sex cells of the male and the female dog, the number of chromosomes has been cut in half by a complex process of cell division. That leaves 39 chromosomes in dogs. When an egg is fertilized, the number of chromosomes goes back to 78. Each of the 39 chromosomes received from the sire has its matching chromosome in the 39 received from the dam.

Classifications of Inherited Characteristics

The laws of inheritance in dogs are the same as in all other organisms, including man. We will now describe the classifications of inherited characteristics.

Recessive. For a recessive characteristic to show in an individual, it must get a gene for a particular trait from *both* parents. If he has it from only one parent, it will possibly be there to show up in some later generation, but it won't be seen in him.

Dominant. This trait will appear whether it comes from only one parent or both. To look at the dog, you could not tell whether the trait came from one or both parents, but it does make a difference in terms of the offspring. From a technical standpoint, if an animal inherits the genes for a particular trait from both parents, he is homozygous—the genes are the same. If the genes are different, he is called heterozygous. The dominant gene is what is seen in the individual. The recessive gene is lurking there to show up, most likely, in a future generation.

Sex-Linked Characters. This term is applied to traits for which the causative gene is carried on the sex chromosomes. The best known example is hemophilia, or delayed blood clotting and excessive bleeding, which occurs in dogs as well as in man and some other animals. It is carried by the female line. The gene is lethal to the male dog, and hence no male offspring with it can survive. (there are numerous other lethal genes, not all of which are sex-linked.)

Polygenic Character. Many inherited conditions are not simply dominant or recessive, are not sex-linked and are not likely to appear in clear cut ratios of normal to abnormal in any particular trait. Intelligence, or a capacity for intellectual activity, some aspects of conformation, resistance to disease and a tendency to certain types of behavior are some of the things that appear to be carried by a group of genes rather than a single gene. Hip dysplasia is often thought of as polygenic, although there are theories and experience suggesting that environmental factors may also play a part. For example, it has been suggested that the chance of dysplasia can be somewhat reduced by using old bathmats or similar washable materials to give traction in the whelping box to keep the pups from slipping and straining their hip joints at an early age. It has also been recommended to keep them rather closely confined rather than let them exercise at will. Also, keep food to about two-thirds of the pup's appetite to prevent too rapid growth. These and other suggestions are all aimed at preventing strain on the bone structure before the muscles are strong enough to keep the bones in their proper position.

The recessive, dominant, sex-linked and polygenic genes are the classes of hereditary factors. Because of many similarities between infant-mother relationships and puppy-bitch relationships, and because in dogs the genetics and the environment can be much better controlled than in humans, several experiments have been conducted.

Experiments with Selected Breeds

Some investigators at the Jackson Memorial Laboratory, Bar Harbor, Maine, who wanted to study the effects of heredity or genetics on behavior in children, developed an elaborate study of dogs. The main part of their testing continued for 13 years. They selected five breeds of dogs of different types, that were well standardized and had been bred over the years or even the centuries for quite different purposes.

Cocker Spaniel. A dog from the Sporting Group, the Cocker Spaniel has been used for over 400 years to search for game birds and then to sit quietly or lie down, or on command to make a retrieve to hand. The Cocker is born sociable.

Beagle. A scent hound, the Beagle has for centuries followed the trail of the rabbit almost entirely independent of any control of the hunter. The Beagle is a natural self-hunter, while the Cocker and others in the Sporting Group work to the commands of the hunter. Training, of course, is involved, but it would be much more difficult, or even impossible to develop a good hunting dog if it were not for the inherited character traits. The Beagle is born sociable, like the Cocker, but also independent.

Basenji. Also in the hound group, the Basenji was selected because it works like a sight hound and is very different from any of the other breeds in the study. It is called the African barkless dog. It can bark, but rarely does. It makes other sounds quite different from those of any other dog. It is a sight hunter and runs in packs. "Its general natural characteristics are aloofness, agility in action and little or no inclination to form close attachments to human beings." They are born wild and aloof, though it is felt that through selective breeding a more sociable strain could be developed.

Wire Fox Terrier. The Wire Fox Terrier was selected because it possesses the aggressiveness to deal with the type of game a terrier is supposed to hunt. It is born aggressive. (When a researcher refers to the term aggressive, there is no connotation of meanness. It means energetic and active, unrestrained.)

Shetland Sheep Dog. This dog was selected to represent the herding dogs. It is very dependent on the will of the shepherd, obeys commands well and is described as almost devoid of ability to make independent decisions. These are valuable traits in its work. It is born with a need of security and approval.

Extensive tests of all kinds were used on puppies of each breed. Cockers and Basenjis were the most extreme in terms of their relations with people. The others showed varying degrees of natural attachment to people. During the study there was close inbreeding to keep the genes the same as when the study started. It is not a comparison of breeds, but a study of inherited characteristics and how they are influenced by environment.

After the traits of each breed were established, there was an elaborate program of cross breeding and testing to get more information about inheritance. A general conclusion was that "each breed has a number of traits of behavior and temperament that combine to make a dog which is highly trainable for the breed's purpose. These traits include such things as aggressiveness, interest in food, quickness to form habits, etc." (Scott, 1965).

In addition to the great importance of inherited characteristics and how inheritance works, the study produced an increased knowledge of the effects of environment. It confirmed what we have known, that environment never has given man or dog any ability he did not inherit. The environment may provide the opportunity for the realization of his full potential, or at the other extreme it may limit development even to the point of complete frustration.

Clarence Pfaffenberger (1965), in his book, *The New Knowledge of Dog Behavior,* gives a striking example of limitation by the environment. A man had a very promising 8-week-old Cocker Spanial puppy. He placed her with a most competent puppy field trial trainer. She became one of America's great Cocker Field Trial Champions. She was bred to a dog with equally good genes from the same strain. The litter of puppies was raised in a stall in a barn. Water and food were placed in the stall once or twice a week. No one was allowed to handle the puppies. At four months of age these puppies would try to bury themselves by digging if approached by anyone. In the field they hunted as if they had been well trained, the result of their heredity, but if anyone spoke to them they would lie flat on the ground and freeze as if paralyzed. Their bodies felt as cold as if they were dead. Beginning at 4 months a great deal of care was given to these young dogs to try to socialize them, but only one ever began to act like a normal dog.

CRITICAL PERIODS OF DOG INTELLIGENCE

Various researchers, such as Konrad Lorenz (1935) with geese and other birds, and John Paul Scott (1965) with a lamb, have found that if an animal associates with a human rather than with its own kind during certain periods in its early life, it will become *imprinted* on that human and reject its own kind. The animal then grows up with characteristics very unlike animals of its own species.

By studying hundreds of puppies that grew up with their mothers, litter mates and human attendants, it was also learned that there are definite crucial periods in their social development. These are the same for all breeds. They are called the *critical periods*. There are four.

Up To Three Weeks

Life is relatively simple for most puppies. They have few challenges, except in competition for the sources of milk. During the first week, they need an environment with a temperature from 85 to 90 degrees and spend their time nursing and sleeping. The mother almost constantly licks them. The first critical period averages about 19½ days, depending on how long the mother carried the pups. With all breeds it is complete by the 21st day. This is the period where the main focus is survival—warmth, food, massage and sleep are the things the puppy needs.

The puppy opens his eyes anywhere from the 11th to the 19th day, the average being the 13th. He seems to use his eyes to a certain extent as soon as they are open, but full visual capacity is not reached until 7 or 8 weeks. Sound reaction appears first at the 21st day. On this day all sense organs seem to be functional and the puppy is no longer dependent on reflex responses to hunger, cold and touch. He now has the ability to see, hear and smell, as well as to taste and feel.

Most puppies start to walk in an unsteady fashion at about the 18th day, and they start playing with one another by chewing on each other in the nest. Up to now they have eliminated by reflex, stimulated by the licking of their abdomens by their mother. Now they have come to the point where they start eliminating independently.

Three to Seven Weeks

At this time, with their senses all beginning to function, there is great development. At 21 days of age they start leaving that part of the nest where they sleep and play to find another corner where they can eliminate. (This is a natural, clean character trait of puppies that makes housebreaking rather simple if started at an early age.) They now go toward objects that attract their attention. They sleep less. They still huddle together to sleep, but each puppy does some private investigating on his own when he is awake. At the beginning of this period his memory span is a few minutes at most, often only a few seconds. But EEG (brain wave) studies show that his brain is beginning to take on its adult form, and at 7 weeks he has an adult brain in terms of capacity, but not the experience.

The environment begins to play an increasingly important part in the development of the dog. The big world about him is opened up to his attention and he needs his mama very much. But he can and should be handled so that he will start to form attachments to people.

During this period he also develops his attachments to other dogs through association with his mother and his litter mates. This is normal and is very important in the well-rounded development of a dog. If they are take away before the end of 7 weeks, they miss some of their canine socialization. Some experience shows that the puppy who does not complete his 7 weeks of canine socialization is often the dog that, when grown, picks fights with all the strange dogs he meets. During this period, playing and often play-fighting begin. In some breeds it can become quite serious fighting, and an order of dominance is begun.

Some experiments showed that when puppies were taken away from their mother and litter mates at 4 weeks and given a lot of human attention, they became very socialized to human beings but did not care about other dogs and some were almost impossible to breed.

A desirable arrangement is to start human socialization, giving the puppy personal affection and a little training beginning at 5 weeks, but leaving him in the litter with the mother for 7 weeks. The age of 7 weeks seems the ideal time for the new owner to take the puppy home. From now to the 16th week, his basic character will be set by what he is taught—or not taught.

Seven to 12 Weeks

This is the period for maximizing their inherited intelligence by giving them as wide a variety of experiences and instructions as their puppy minds and emotions are capable of absorbing. A daily 15 minute training session during this period seems desirable and speeds up the learning process. Their attention span is short. About three times of the same thing is generally all that is useful.

In summarizing some of their research, Dr. Scott said, "The evidence from puppies is that they have a short period early in life when positive social relationships are established with members of their kind and after which it becomes increasingly difficult or impossible to establish them. The same applies to their relationship with human companions. The period in puppies when we can best socialize them and begin their training is in the period of 5 weeks to 12 weeks of age."

Up to 7 weeks of age, a certain amount of trying-out dominance is good for a puppy. It causes him to have respect for other members of his species and also develops something of an attachment to his own kind. But to leave him in his kennel run with his litter mates until he is 16-weeks-old, without giving him special individual attention entirely away from his litter mates, is depriving him. This includes basic training and affection. Without this, he will either be a life long bully or an underdog. It will make him poor material for any training and will limit his ability to adapt to human companionship.

Twelve to 16 Weeks

This has been called the *age of cutting*. If allowed any freedom, the puppy cuts his mother's apron strings and declares his independence. He wanders away alone or with a companion, gets into mischief and cuts his teeth both literally and figuratively. At this age he can still be socialized to human beings. He can be started in training. But it will never make up for anything lost through neglect in earlier training. This is the time when man and dog decide who is boss. Serious training can and should be started—a transition from play training to disciplined behavior.

A puppy who has had no socializaton before he is 16-weeks-old has little chance of becoming the sort of dog that anyone would want as a companion. He needs to be taken entirely away from the other dogs and have a pleasant session of getting acquainted or of training. Even two 20 minute sessions a week, on a regular schedule, should accomplish this.

The time is short—up to the age of 16 weeks. Once it is gone, this time can never be retrieved. Puppy raisers should employ it wisely. Puppy kindergarten training is a big help in providing socialization with different people and different dogs. Walks through a shopping center, short drives in the car—all contribute to giving the puppy new and different experiences.

Dr. Scott states the importance of the critical periods this way, "As different as are the inheritances of different breeds of dogs, all when given proper socialization from 3 weeks to 16 weeks of age, will reach a satisfactory level of behavior. Social relations are formed through the process of learning. They begin at a point where the first capacity for learning appears (at 21 days). It is important to remember that, while previous learning may be altered by subsequent learning, subsequent learning will never obliterate previous learning."

Other factors that influence a dog's best use of his intellectual potential are his physical health and the suitability of the environment you give him including space, temperature, security, stimulation, consistency of treatment and respect for him as a member of the household without pampering. All these and others have a bearing on how intelligently the dog functions. Dogs are highly adjustable. They can adapt to quite restricted conditions, but such conditions do not contribute to their intellectual development.

One animal psychologist has the theory that a dog (and man, too) accumulates a sort of file of thought patterns on which he can draw to solve puzzles or problems with which he is confronted. A dog kept in a kennel has nowhere near the opportunity to accumulate patterns that the farm dog has, or a dog in some other rich and varied environment. If we want our dogs to make the best possible use of their inherited intellectual potential, use the critical periods wisely by playing, training and working with them. Make their young lives as rich in different experiences as possible.

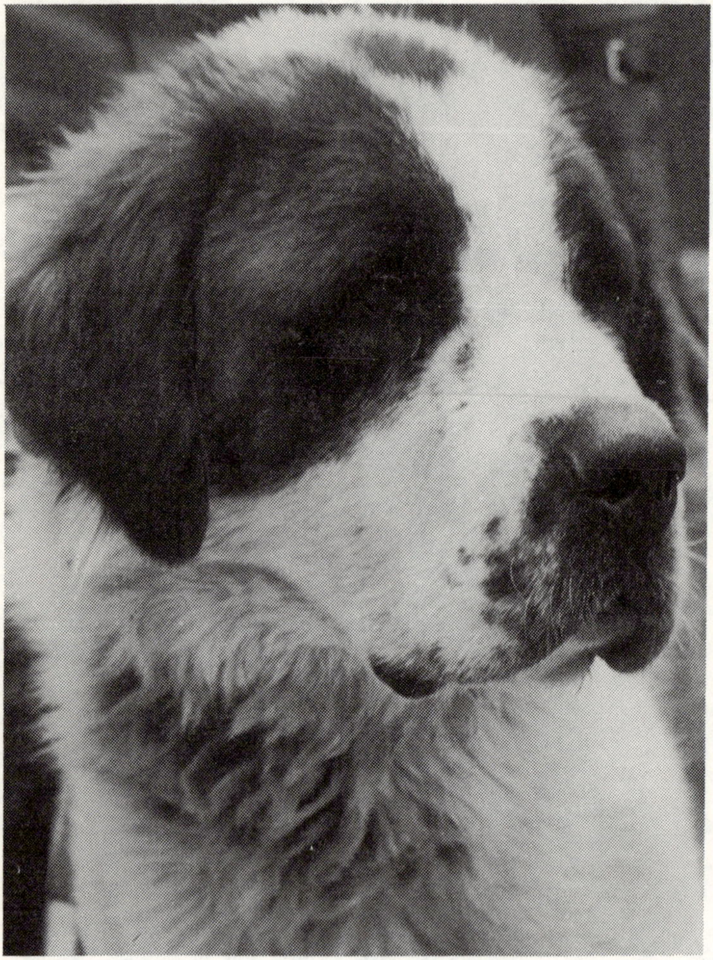

Chapter 8
General Aspects of Test Administration

The dog must be examined by a person who has a lot of patience, likes dogs and is able to relate to them (Fig. 8-1). You cannot become disgusted, fuss at the dog, speak loudly to him or scare him. The examiner must remain calm at all times and try to get the dog thinking that the test is like playing games with him. The examiner and the dog should have fun together.

Tests should be presented in the sequence given in this book in order to alternate those with and without food rewards. In this way the dog is less apt to be satiated before the end of testing. Dogs should be reasonably hungry at the time of testing. The length of time during which food is withheld will depend on the age of the dog. For example, puppies 6 months of age or less should not have eaten for about four hours.

Use an average size room in the home. The dog being tested should be the only dog present. The area should be free from distractions to the dog. For example, the television should not be on during testing and there should not be intriguing items lying around for the dog to pick up and play with.

Test items should be arranged so that a minimum of time is spent in getting the items required for each test and so that other items do not provide a distraction. A small table, desk or counter is desirable for arranging the items

Fig. 8-1. This examiner obviously likes dogs and is able to relate to them.

for each test. It will also be helpful for recording scores and notes.

THE EXAMINER

When the examiner is first exposed to the dog, it is often helpful to have the owner call the dog, take the examiner by the arm and say to the dog, "This is Fred. He's your friend." The owner should pat the dog and say, "Good Fritz" and pat the examiner on the arm saying, "Good Fred." This may be repeated several times. The examiner then can pat the dog and pat the owner to strengthen the feeling that all three are friends.

Fig. 8-2. The examiner should establish a friendly relationship with the dogs.

The examiner should establish a friendly relationship with the dog as soon as possible (Fig. 8-2). The way he does this and the amount of time required depends on the dog. Some are naturally aloof and require quite a lot of time to warm up and make friends. Others love everyone and everyone is automatically their friend. Then there are those dogs that are suspicious and even hostile to strangers. It usually requires a great deal of time to overcome this attitude. It is often good to stoop or kneel in order to get your face almost on the same level as the dog's face (Fig. 8-3). Don't bend or tower over a dog. The idea is to get yourself on the same plane as the dog. Extend the back of your hand toward the dog's nose and let him sniff it. Don't use an outstretched hand to pat him on the head. This can be interpreted as a threatening gesture. After the dog has sniffed your hand, remain in a relaxed position speaking to him softly and calling his name. Avoid looking at the dog in the eyes for more than a second or so. If you do, this may create what is called the *staring syndrome*. This is often disturbing to dogs. Watch him through the corner of your eye rather than looking at him directly. The reason for this is that dogs in the wild that were being stalked by other animals were usually stared at by predators before the predators pursued. This primitive experience seems to have been carried over to domestic dogs. They can be quite disturbed by having even a friendly person stare at them intently.

Fig. 8-3. The examiner should stoop or kneel in order to get her face on the same level as the dog's face.

TEST SEQUENCE

This book contains a total of 16 tests. Ten are in the regular test battery and six are alternate tests. Those in the regular battery have been selected because they generally have a higher interest and usefulness in examining a dog's intelligence. If for some reason one or more of these don't seem suitable for a particular dog, alternative tests can be substituted. The tests in the main battery are arranged in approximate order of expected responsiveness on the part of the dog. Most dogs will respond positively to the first few tests even if they tend to get bored and apathetic as the test progresses. Many of the test items use food as a reward. These have been alternated with non-food tests to reduce the danger of a dog becoming satiated and consequently uninterested in working for food.

The examiner must use his judgment in regard to the time allowed between tests. Some dogs are willing and eager to go through the whole battery without any breaks, while others want some play time after each test. In some instances, particularly with very young puppies, two testing sessions may be advisable because of their short attention span.

TEST PREPARATION

The examiner should familiarize himself thoroughly with the test items and the procedures for each test. He should have a score sheet and a pad for notes immediately at hand. The material should be organized in sequence for each test so that there is minimal delay in moving from one test to the next.

A number of the tests require tidbits as motivating elements or rewards. Since dogs' tastes vary, it is well to have a variety of treats. In our experience the best motivating treats are small balls of raw hamburger or little pieces of cooked liver. In some instances the dog owners might be invited to bring some treats that their dog particularly likes. Unless the treat is something the dog really likes so that he is willing to work to get it, the test becomes invalid.

At times the examiner may experience difficulty in getting any response from the dog on a particular test. In this case it is wise to move on to the next test which the dog may find more interesting and then later return to the test which was skipped.

Although these are basically fun tests and do not put the dog in a mental category which will follow him for life, they do give an indication of his alertness, responsiveness and adaptability to various kinds of situations. They are particularly useful in testing puppies from the same litter that are 8 weeks or older. They are also useful when testing dogs with similar life experiences. In such cases the scores of the different animals will carry a substantial significance. A breeder may well want to use them as one factor in deciding which dog to keep.

In some cases the dog may be so closely attached to the owner that it assumes the role of a spoiled, dependent child. If so, a valid evaluation will not be obtained when the owner is present. When such a relationship exists, the owner should be asked to leave the examining room.

TIME LIMIT

If the examiner feels that it is useful to allow more time than indicated in order to get a better sample of behavior and more information about the dog's approach to problems, the test may be continued as long as it seems useful do to so. For example, in Test Numbers 6 where the dog is trying to extract food from under a piece of furniture and is working at it vigorously at the end of the time limit, he should be permitted to continue so that the examiner will have an idea of his persistence and degree of frustration tolerance. The examiner should discontinue the activity after all useful information has been gained or when it no longer seems useful in understanding the dog.

SCORING

The scoring of some of the tests is very clear. The dog earns a score from zero to four. These scores are clearly defined and observable on an objective basis. There are other tests, however, where interpretation on the part of the examiner is required. For example, in Test 7 it is often difficult to tell whether a dog attended to a sound or not. The examiner must make a judgment regarding the score. With testing experience this becomes easier and the examiner is

Table 8-1. Score Sheet.

TEST NUMBER AND DESCRIPTION	TIME	SCORE
1. Towel over head 2. Food wrapped in paper 3. Rubber band on muzzle 4. Finding food in room 5. Paper bag over head 6. Finding food under furniture 7. Response to noise 8. Food under can 9. Food concealed in hand 10. Verbal responsiveness **ALTERNATIVE TESTS** 11. Transparent barrier 12. Food covered by towel 13. Select food from one of two cans 14. Select food from one of three cans 15. Removing cord around head 16. Paper barrier		

better able to be consistent in his judgments regarding what might appear to be ambiguous performances. After testing each dog the examiner should review the score sheet (Table 8-1) to see that all entrees have been made legibly. He should also enter any special comments regarding the performance of the dog. If this is not done immediately, there is danger of confusing the performance of one dog with another.

TEST RESULTS

After an examination one should always review the score sheet and form an opinion as to whether or not the score is consistent with the impression the dog gives and the dog's history. If there appears to be any marked discrepancy between the dog's score and the score he was expected to get on the basis of history and observation, the examiner should review the test results for possible error. If none is found, the examiner might want to make an appointment for re-examination a week or so later. When giving the owner a report of the test results it should be made clear that different breeds of dogs do better on some tests than others. The standardization of these tests has been very limited to date. The IQ actually does little more than say that Fritz is smarter than average, about average or not quite up to the average.

OBTAINING DIQ (DOG INTELLIGENCE QUOTIENT)

The examiner might obtain a *DIQ* with these tests. Such a score would obviously have no validity except for purposes of comparison among several dogs with which you may be working (Fig. 8-4).

Intelligence in humans does not proceed by equal amounts through the life span. At early age levels the mental growth is rapid. At the upper ages the level of mental functioning declines slightly. For most intelligence scales the difference between human ages 15 and 25 are negligible. Above that age there is a slight gradual decline. We know from human testing that the age of greatest mental ability is between 22 and 25 years. Beginning at about age 25 there is generally a slow gradual decline.

In determining the IQ of a dog we must take into account chronological age which reflects mental growth. A 3-month-old puppy generally does not have the mental capacity of a 5-year-old dog because the latter will have had much more opportunity for training, improving skills and dealing more effectively with the environment. There is virtually no information on the intellectual growth and development of dogs that we could find, and in developing our measurement, we will make some assumptions.

The first is that after a dog passes early puppyhood, he gains in intelligence up to about age 3. This is providing he has had good enriching experiences. After that intelligence remains relatively constant for a few years.However, at about age 6, it gradually declines in the same way that human intelligence gradually declines. For lack of better data the assumption is made that after a dog reaches adulthood one year of his life is equal to about seven years of human life.

Another assumption is that the growth and development of dog intelligence follows approximately that of human intelligence. Accordingly, in terms of chronological age a dog between the age of 3 and 3½ years is at his peak of intellectual ability. Since decline in human intelligence is at about the 35-40 year range, we have used age 6 as the point at which dog's mental performance would be expected to show a slight decline.

Table 8-2. Scoring a Dog's IQ.

DOG AGE	MULTIPLY TEST SCORE BY
Under one year	—
From 1 up to 2 years	6
From 2 up to 3 years	5
From 3 up to 7 years	4
From 7 up to 9 years	5
From 10 years and older	6

Dogs between the ages of 3 and 6 are used as our base in determining the dog's IQ. To obtain IQs for dogs between ages 3 through 6 we suggest multiplying the test score by 4. Thus, a 5-year-old dog with a test score of 30 would receive an IQ of 120 (30 × 4). For dogs ages 7 up to 9 we suggest multiplying the test score by 5 to allow for the decline in

Fig. 8-4. A DIQ score has no validity except for purposes of comparison among several dogs with which you may be working. (Photo courtesy of Allied Food Inc., Wayne Dog Food.)

mental abilities. Thus, an 8-year-old dog with a test score of 25 would receive an IQ of 125. For dogs 10 years and older we suggest multiplying the test score by 6.

We must also take into account dogs younger than 3 years old to allow for their incomplete mental development. We suggest that for dogs from 2 up to 3 years the test score be multiplied by 5. Thus, a 2½-year-old dog with a test score of 20 on the test items would receive an IQ of 100. For dogs from 1 up to 2 years we suggest that the test score be multiplied by 6. Thus, a dog 1½-years-old with a score of 20 would receive an IQ of 120.

This test is not considered valid for dogs under one year of age, but the test items are helpful in studying litters of puppies from an age even as young as 5 weeks. The test also provides good training, enriching and socializing experiences for the puppies. A recapitulation of the procedure to obtain a dog's IQ is found in Table 8-2.

Results in terms of DIQ may be classified into rough groups as follows:

- 130 and above   Very Superior
- 120 to 129   Superior
- 110 to 119   High Average
- 90 to 109   Average
- 80 to 89   Low Average
- 70 to 79   Borderline
- 69 and below   Mentally Deficient

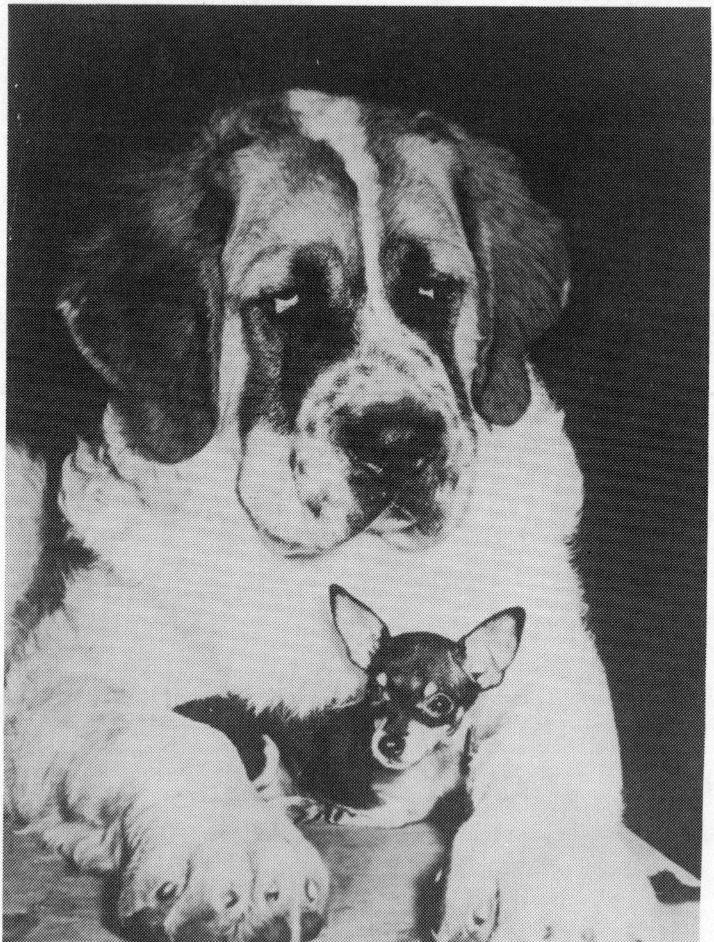

Chapter 9
Administering and Scoring the Tests

An important selection criterion of test items is the availability of testing materials in a home setting. All of the materials in the test are probably readily available around the house. The materials can be put away in a convenient box. The examiner should first place the test items on a small table, desk or counter where he can easily see them. He should then arrange the test pieces in the order in which he is going to give the test. He can then quickly lay his hands on the test item when it is next needed. When each particular test is completed, that test item should be put away in the box so that it does not get mixed up with the test items yet to be used.

Materials for the regular 10-test battery (Fig. 9-1) include:

- Tidbit
- Towel, small blanket or heavy cloth
- Paper napkin or newspaper
- Rubber band
- Paper bag
- Pebbles, nails or coins
- Tin can

Additional materials for the six alternative tests (Fig. 9-2) include:

- Plate glass or plexiglass

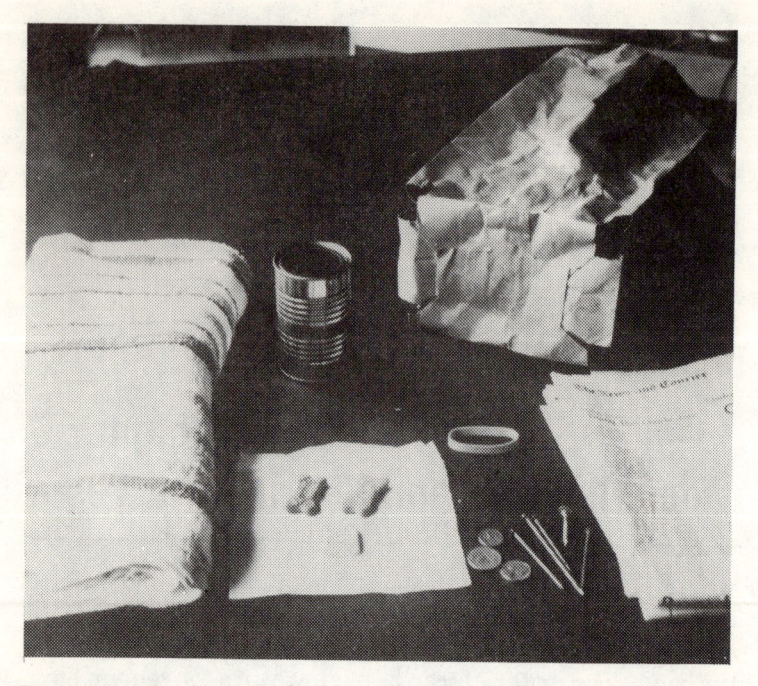

Fig. 9-1. Complete set of materials for administering the regular 10-test battery.

- Two additional tin cans
- Cord, yarn or ribbon
- Rubber ball
- Double-page newspaper
- Towel

TEST NUMBER 1

Measures

Common sense and judgment.

Materials

Towel, small blanket or heavy cloth.

Procedure

Take a large bath towel, small blanket or heavy cloth of similar size and present it to the dog to let him smell it. Then with a quick motion throw it over his head so that the head is

114

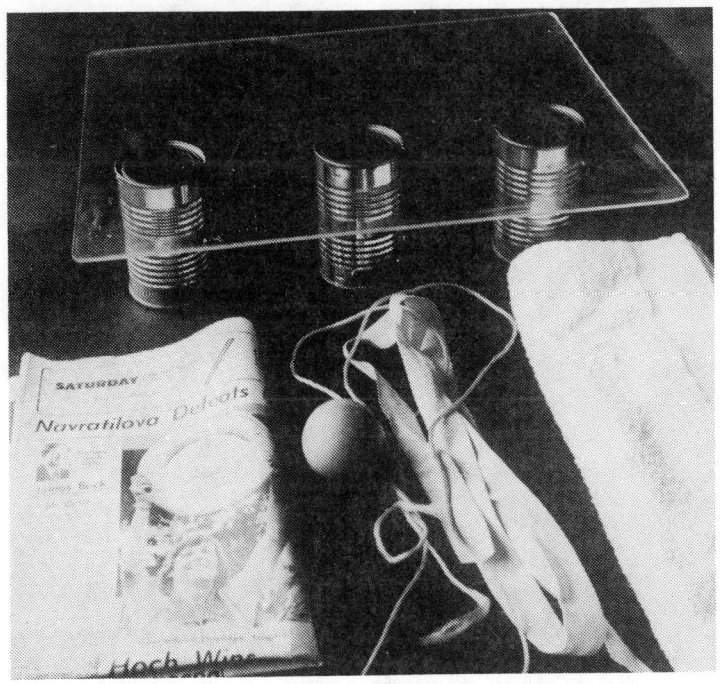

Fig. 9-2. Materials for administering the six alternative tests.

completely covered. Time him to see how long it takes to get his head free (Figs. 9-3, 9-4, 9-5, 9-6, 9-7, 9-8 and 9-9).

Scoring

4 points: The dog frees himself in 15 seconds or less.
3 points: The dog frees himself from 16 to 30 seconds.
2 points: The dog frees himself from 31 to 60 seconds.
1 point: The dog takes more than 60 seconds to free himself.
0 points: The dog gives up and does not free himself.

TEST NUMBER 2

Measures

Problem-solving ability.

Materials

Tidbit, napkin or small piece of newspaper.

Fig. 9-3. The examiner has just thrown a towel over Charlie's head. He is now timing to see how long it will take Charlie to remove the towel.

Fig. 9-4. This young examiner has just thrown a towel over Daisy Mae, a 7-month-old Basset Hound.

Fig. 9-5. A young woman is examining Ralph, a 2½-year-old Bulldog. The towel has just been put on the dog and he is beginning the struggle to free himself.

Fig. 9-6. A young examiner has just thrown a towel over Belle, a 1-year-old female Bulldog.

117

Fig. 9-7. Butch, a 2-year-old black Labrador Retriever is being tested. He is making good progress in getting the towel off his head.

Fig. 9-8. A 5-year-old girl is testing Abbie, an 8-month-old mixed breed. Abbie is almost completely covered by the towel.

Fig. 9-9. Abbie has succeeded in getting her head uncovered and will soon free herself entirely.

Procedure

Permit the dog to see and smell a piece of food. While the dog is looking, wrap the food in a paper napkin or small piece of newspaper in a loose bundle and throw it on the floor. Observe the dog's behavior as he tries to obtain the food (Figs. 9-10, 9-11 and 9-12).

Scoring

4 points: With paws or mouth the dog gets the food within one minute.
3 points: Same as above but requires from two to three minutes.
2 points: Same as above but requires more than three minutes.
1 point: The dog tries to obtain the food but soon gives up.
0 points: The dog makes no effort to get the package.

TEST NUMBER 3

Measures

Common sense and judgment.

Fig. 9-10. Food has been placed in a paper napkin. Charlie is sniffing it and will soon get it out of the paper. The examiner is timing him.

Fig. 9-11. This young examiner is testing Schatze, a 4-month-old Dalmatian. Schatze struggles to get the napkin open.

Fig. 9-12. Schatze has opened the napkin and is getting the food.

Materials

Wide rubber band slightly smaller in diameter than the dog's muzzle so that it will exert a little pressure.

Procedure

While the handler holds the dog, the examiner places the rubber band half way up the muzzle of the dog. At the same time the handler releases the dog so that he is free to work at removing the rubber band (Figs. 9-13, 9-14, 9-15 and 9-16).

Scoring

    4 points: The dog removes the rubber band within 5 seconds.

    3 points: The dog removes the rubber band within 15 seconds.

    2 points: The dog removes the rubber band within 30 seconds.

    1 point: The dog paws its muzzle or rubs the muzzle on the ground but does not remove the rubber band after one minute.

    0 points: The dog makes no effort to remove the rubber band after one minute.

Fig. 9-13. The examiner has placed a rubber band around Charlie's muzzle. Charlie is trying to remove it. The examiner again is timing him.

Fig. 9-14. A rubber band has been placed on the muzzle of a 6-year-old Beagle, Happy, by the examiner. Happy is using both his front paw and the ground in order to remove the rubber band.

Fig. 9-15. The rubber band is successfully removed and Happy is happy.

Fig. 9-16. The examiner has just given the rubber band test to Molly, a 1½-year-old Golden Retriever. Molly has just taken the rubber band off her muzzle and is relaxing.

Fig. 9-17. Fritz was shown a tidbit. While he was looking on, the examiner placed the tidbit in the corner of the room. Fritz was taken out of the room, led around in a small circle and returned to the center of the room and released from the leash. He moved toward the corner of the room to find the tidbit. The examiner observes his behavior in finding the food.

TEST NUMBER 4

Measures:

Problem-solving ability and concentration.

Materials:

Tidbit.

Procedure:

Select an average size room which contains only a small amount of furniture or other barriers. Take the dog in the room on a leash. Show him a desired tidbit and while one person is holding the dog in the center of the room, have the examiner place the tidbit in one of the corners of the room. Be sure the dog sees this being done. Take the dog from the room, lead him around in a small circle and bring him back to the center of the room. Release him from the leash and time his progress (Figs. 9-17 and 9-18).

Scoring

4 points: The dog goes immediately to the tidbit.
3 points: The dog finds the tidbit by only systematically sniffing around the edge of the room.
2 points: The dog displays somewhat random behavior and appears to find the tidbit.

1 point: The dog tries to find the tidbit in random fashion but does not succeed.
0 points: The dog makes no effort to find the tidbit.

TEST NUMBER 5

Measures:

Common sense and judgment.

Materials:

Paper bag.

Procedure

Select a paper bag which will slip comfortably over the dog's head and yet is reasonably tight. Cut two or three holes about 1 inch square at the bottom of the bag to provide ample air for breathing. While the handler holds the dog in a standing position, the examiner should slip the bag over the dog's head and start timing. At the same time the handler releases the dog so he is free at removing the bag (Figs. 9-19 and 9-20).

Scoring

4 points: The dog removes the bag within 5 seconds.
3 points: The dog removes the bag within 15 seconds.

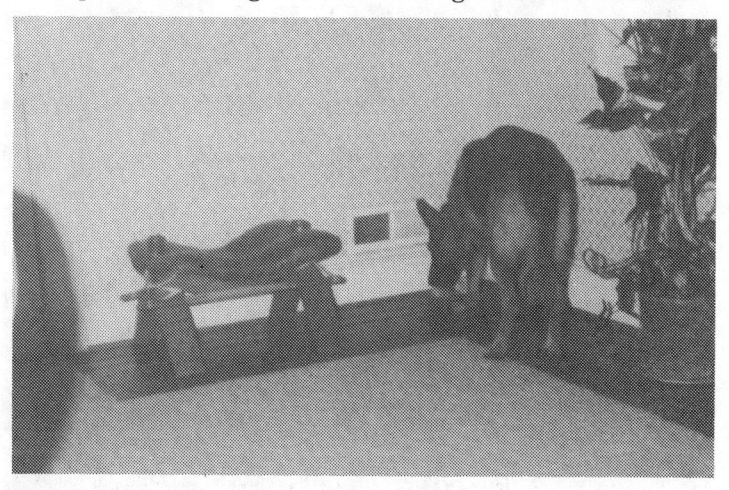

Fig. 9-18. Fritz finds the tidbit in the corner of the room.

Fig. 9-19. The examiner has placed a paper bag with holes in it to provide ventilation over Charlie's head. He is timing to see how long it will take Charlie to remove it.

Fig. 9-20. This young examiner is testing a 2-year-old mixed breed, Herman. The bag has just been placed on Herman's head and he is starting to raise his left front leg to get it off.

2 points: The dog removes the bag within 30 seconds.

1 point: The dog paws at the bag and makes slight effort but does not remove the bag or the dog removes the bag within one minute.

0 points: the dog makes no effort to remove the bag after one minute.

TEST NUMBER 6

Measures

Problem-solving ability and concentration.

Materials:

Tidbit.

Procedure

Permit the dog to see and smell a piece of food, such as a piece of meat. While the dog is looking, place the food under a piece of furniture, such as a sofa or chair. The furniture should be low enough to that he cannot get his head under but high enough to permit him to paw it out. The dog is then encouraged to retrieve the food. Observe the method that he uses to get it and how long it takes (Figs. 9-21, 9-22, 9-23 and 9-24).

Scoring

4 points: The dog quickly sizes up the situation, paws and retrieves the food within one minute.

Fig. 9-21. The examiner has placed a tidbit under a piece of furniture in full view of Charlie but out of his easy reach. Charlie is looking for the tidbit.

Fig. 9-22. Charlie is now sniffing the food.

Fig. 9-23. Charlie is pawing at the tidbit.

3 points: The dog sizes up the situation and retrieves the food within three minutes.

2 points: The dog does not use his paw, tries to use his muzzle but fails to get the food.

1 point: The dog does not use his paw, only sniffs and puts out no further effort.

0 points: The dog makes no attempt to retrieve the food.

TEST NUMBER 7

Measures

Attention.

Materials

Tin can and pebbles, nails or coins.

Procedure

Without the dog noticing, place a can with a few pebbles, nails or coins (something to make a noise) behind your back. Stand about 6 feet from the dog and rattle the can vigorously for a few seconds. Observe his response to the noise and exploratory behavior (Figs. 9-25, 9-26 and 9-27).

Scoring

4 points: The dog becomes instantly alert and starts in search of the source of the noise.

3 points: The dog is alert, becomes mildly startled but

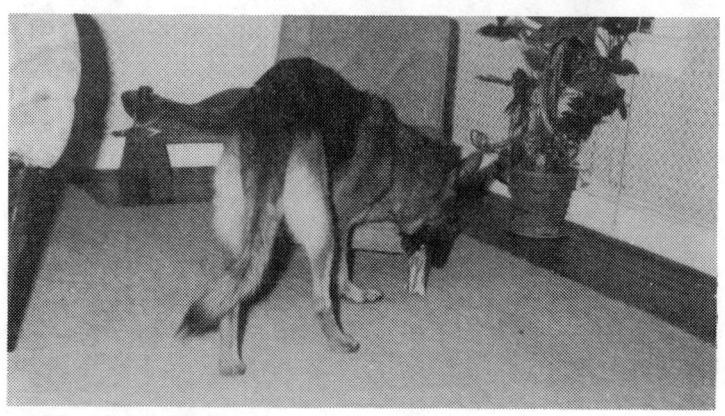

Fig. 9-24. Charlie retrieves the food.

Fig. 9-25. The examiner places a can with coins in it to make a noise behind his back, while the dog watches.

Fig. 9-26. The examiner stands about six feet from the dog and rattles the can vigorously.

Fig. 9-27. The dog seems instantly alert and starts in search of the noise.

recovers and starts to investigate the source of the noise.
2 points: The dog is alert but will not investigate the source of the noise.
1 point: The dog is alert but backs away, has an avoidance response and does not investigate.
0 points: The dog shows fear, cringes and moves away from the examiner.

TEST NUMBER 8

Measures

Problem-solving ability.

Materials

Tidbit, tin can.

Procedure

The handler is not needed in this test. The examiner places the can upside down on the ground or on the floor. He holds a tidbit in his hand and lets the dog sniff it. He places the tidbit under the can and starts timing. He encourages the dog to get the tidbit (Fig. 9-28 and 9-29).

Fig. 9-28. The examiner holds a tidbit and lets Charlie sniff it. Then he places the tidbit under the can and starts timing to see how long it will take Charlie to retrieve it.

Fig. 9-29. A young woman is starting to examine Crissie, a 10-month-old Doberman Pinscher by placing a tidbit under a tin can. She is showing Crissie the food and is about to place the can over it.

Scoring

4 points: The dog retrieves the tidbit within 5 seconds.
3 points: the dog retrieves the tidbit within 15 seconds.
2 points: The dog retrieves the tidbit within 30 seconds.
1 point: The dog paws at the can and gets the tidbit within one minute.
0 points: The dog makes no effort to retrieve the tidbit.

TEST NUMBER 9

Measures

Problem-solving ability and concentration.

Materials

Tidbit.

Procedure

The handler holds the dog on a leash. The examiner places a tidbit in each fist but doesn't let the dog see the food. He places his fists in front of him. He then turns the right fist open with the palm up and shows the tidbit to the dog for five seconds. Then he closes the right hand and places both fists behind his back for 15 seconds. He calls the dog and encourages him to get the tidbit (Fig. 9-30).

Scoring

4 points: The dog comes directly to the right hand within five seconds.
3 points: The dog comes to the right hand within 15 seconds.
2 points: The dog comes to the right hand within 30 seconds.
1 point: The dog goes first to the left side and left hand but then shifts and goes to the right hand within one minute.
0 points: The dog makes no effort to retrieve the tidbit.

TEST NUMBER 10

Measures

Receptive language (vocabulary).

Fig. 9-30. The examiner placed a tidbit in each fist but did not let Charlie see the food. Both fists were placed in front of Charlie. Then he turned the right fist and opened it with the palm up. He showed the tidbit to the dog for 5 seconds. Then the examiner closed the right hand and placed both fists behind his back for 15 seconds. He then called Charlie and urged him to get the tidbit.

Materials

None.

Procedure

Part I: Two people are needed for this test. One holds the dog by a short lead and the other goes about 10 feet away and calls the dog's name (example: "Fritz"). If the dog goes to the person calling him or shows some responsiveness or tendency to go to him on the first call, go to the second part of this test. If the dog does not respond, give him a second trial.

Part II: The dog is held by the collar and the caller is stationed about 10 feet away and calls, "Fritz, come!" Simultaneously, the person holding the dog releases the collar. If the dog goes directly to the caller, go on to part III. If the dog goes to the caller but does not do so directly, or if he does not respond, give the dog a second trial.

Part III: The dog is taken close to the handler who gives the command, "Fritz, sit." If the dog fails to sit, give him one more trial (Figs. 9-31, 9-32 and 9-33).

These three words (dog's name, come and sit) may be regarded primarily as examples. Dogs are exposed to many different verbal experiences through training for work and through the teaching of tricks. Also, the dog hears a lot of language in the household. In that case, only the owner can know what words the dog is likely to understand and respond to.

For example, we have a friend whose dog Gipsy responds to the command, "Say your prayers." Gipsy responds by putting her front paw on the edge of a low chair and bowing her head. She remains in that position until she is rewarded. It is suggested that the owner prepare a list of words which he feels would be fair for the dog to be tested with. Score the dog on the basis of the words to which he responds to correctly on the first trial.

Scoring

Part I

1 point: The dog looks at the caller and shows some respon-

Fig. 9-31. Fritz responds to the verbal command, "Sit."

Fig. 9-32. Charlie is sitting in response to the first verbal command, "Sit." He receives two points on this part of the test.

Fig. 9-33. The examiner is testing Bruno, a 2-year-old Bullmastiff on receptive language. Bruno has responded immediately to the command, "Sit."

siveness indicating that he knows his name.

½ point: The dog is able to do the above on the second trial.

0 points: The dog fails to respond to his name.

Part II

1 point: The dog goes directly to the caller or shows some responsiveness or tendency to go to him on the first call.

½ point: The dog is able to do the above on the second trial.

0 points: The dog fails to respond to the command.

Part III

2 points: The dog obeys the command on the first trial.

1 point: The dog obeys the command on the second trial.

0 points: The dog fails to obey the command on both trials.

To score this test, add the points from Parts I, II and III. Maximum score is 4 points.

TEST NUMBER 11

Measures:

Problem-solving ability and concentration.

Materials

Piece of plate glass or plexiglass varying in size from 8 by 12 inches to 14 by 16 inches, depending on the size of the dog.

Procedure

The piece of glass is held on its short edge about 3 feet in front of the dog who is held on a leash by the handler. The examiner places a tidbit behind the glass with the dog looking on. The dog is taken off the lease and the examiner urges the dog to get the goody. (Figs. 9-34 and 9-35).

Scoring

4 points: The dog goes around the glass immediately and gets the tidbit.

3 points: The dog goes directly toward the glass, touches it with his paws or nose but then goes around the glass. He gets the tidbit within 15 seconds.

Fig. 9-34. The examiner has just placed a tidbit behind the plexiglass as Charlie looks on. The examiner is urging Charlie to get the food.

Fig. 9-35. An examiner is testing Esther by placing food behind a sheet of plexiglas. Esther observes with interest and will soon succeed in getting the food.

2 points: The dog goes directly toward the glass, is blocked in its track but goes around the barrier within 30 seconds.

1 point: The dog goes to the glass and paws at the glass, but goes around the barrier and retrieves the tidbit within one minute.

0 points: The dog does not respond to the food behind the glass and makes no effort to get it.

TEST NUMBER 12

Measures

Problem-solving ability.

Materials

Towel.

Procedure

The handler is not needed in this test. The examiner shows the dog a tidbit for about five seconds. He places it on the floor and at the same time throws a towel over it and urges the dog to retrieve it (Figs. 9-36 and 9-37).

Scoring

4 points: The dog retrieves the tidbit within 15 seconds.
3 points: The dog retrieves the tidbit within 30 seconds.
2 points: The dog retrieves the tidbit within one minute.

1 point: The dog removes the towel and exposes the food but fails to return and take the food.

0 points: The dog makes no effort to retrieve the food.

TEST NUMBER 13

Measures

Problem-solving ability and concentration.

Materials

Two tin cans about the size of an 11 ounce soup can and a tidbit.

Fig. 9-36. The examiner showed Charlie a tidbit for about 5 seconds. He placed it on the floor and covered it with a towel and urged Charlie to retrieve it. Charlie is sniffing the food. The examiner timed 'the dog' to see how long it took him to retrieve the food.

Fig. 9-37. A young examiner has placed food under a towel. Butch, a 2-year-old black Labrador Retriever has pushed the towel away and is eating the food.

Procedure

The examiner places the cans about 1 foot apart. The handler holds the dog about 5 feet away from the cans. The examiner places a tidbit under the can to the examiner's right without the dog noticing this. The examiner sits about two feet behind the cans facing the dog. The examiner shows the dog a tidbit and then quickly places it under the can to the examiner's left. After a wait of 15 seconds the handler releases the dog (Fig. 9-38).

Scoring

4 points: The dog goes directly to the correct can within 5 seconds.

3 points: The dog goes to the correct can within 15 seconds.

2 points: The dog goes to the correct can within 30 seconds. Or he goes to the wrong can but then shifts and goes to the correct can within 30 seconds.

1 point: The dog goes to the correct can within one minute.

Fig. 9-38. The examiner places the tidbit under one can and times Charlie on how long it takes him to retrieve it.

Fig. 9-39. The examiner showed Charlie a tidbit. Then, he quickly placed it under the middle can. After waiting 30 seconds, the handler released Charlie, and the examiner timed him to see how long it took him to retrieve the food.

0 points: The dog makes no effort to go to the correct can or goes to the wrong can but does not shift to the other can.

TEST NUMBER 14

Measures

Problem-solving ability and concentration.

Materials

Three tin cans about the size of an 11 ounce soup can and a tidbit.

Procedure

Place the three cans about 6 inches apart. The examiner places a tidbit under each of the end cans before the test begins without the dog seeing this. The handler holds the dog about 4 feet away from the cans. The examiner sits about 2 feet behind the cans facing the dog. The examiner shows the dog a tidbit and then quickly places it under the middle can. After waiting 30 seconds the handler releases the dog (Fig. 9-39).

Scoring

4 points: The dog goes directly to the middle can and retrieves the food within five seconds.

3 points: The dog goes to the correct can within 15 seconds and retrieves the tidbit.

2 points: The dog goes to the correct can within 30 seconds and retrieves the food or goes to the wrong can but then shifts and goes to the correct can and retrieves the tidbit within 30 seconds.

1 point: The dog goes to the correct can within one minute and retrieves the food despite going to one or both of the wrong cans first.

0 points: The dog makes no effort to go to the correct can or goes to the wrong cans.

TEST NUMBER 15

Measures

Common sense and judgment.

Materials

A piece of soft heavy cord, yarn or ribbon from 3 to 4 feet long or long enough to encircle the dog's head with each end touching the floor.

Procedure

The examiner makes a loop in the middle of the cord with a single knot at the top. While the dog is being held, the examiner slips the loop over the dog's head with the top of the loop just below the dog's eyes and the ends go down to the ground. It is not pulled tight enough to be uncomfortable. The dog is released and timed to see how long it takes him to get the cord off (Fig. 5-40).

Scoring

4 points: The dog removes the cord within 5 seconds.
3 points: The dog removes the cord within 15 seconds.
2 points: The dog removes the cord within 30 seconds.
1 point: The dog paws at the cord but does not remove it.
0 points: The dog does not make any effort to remove the cord within one minute.

Fig. 9-40. Charlie has had the cord tied lightly around his muzzle. He seems to be considering what to do.

TEST NUMBER 16

Measures

Problem-solving ability.

Materials

Double-page newspaper and rubber ball.

Procedure

This test should be done in a room where there is a rug so that the dog has traction. A double-page newspaper is opened and fastened with tape to both sides of a doorway, archway or similar opening. The handler holds the dog on a leash about 4 feet on one side of the barrier. The examiner places himself in a kneeling position about 5 feet on the other side. The examiner bounces the ball several times—high enough for

144

Fig. 9-41. Fritz is approaching the newspaper barrier as a start toward getting the ball which the examiner has just bounced to attract his attention. The examiner is timing his performance and will score him.

the dog to see it—and encourages the dog to come to him. At the same time the handler releases the dog (Fig. 9-41).

Scoring

>4 points: The dog goes immediately over or under the paper barrier.
>
>3 points: The dog goes over or under the barrier within 15 seconds.
>
>2 points: The dog goes over or under the barrier within 30 seconds.
>
>1 points: The dog goes over or under the barrier within one minute.
>
>0 points: The dog makes no effort or goes over or under the paper beyond one minute.

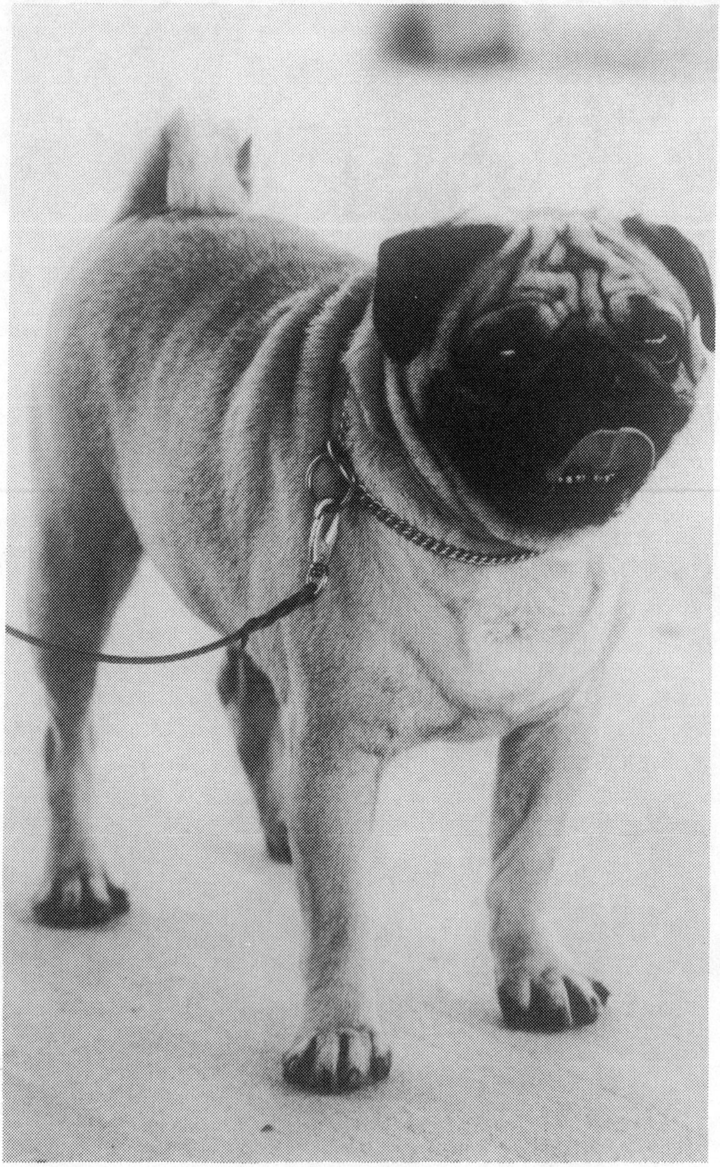

Chapter 10
Increasing Your Dog's IQ

As reported in chapter 6 various studies have shown ways of enhancing intelligence. With puppies a very important method used by training organizations is the puppy kindergarten where the young animal is exposed to various people, in addition to his owner, to other animals and to a wide variety of toys, sounds, objects, etc. For the single dog owner or the owner of an older dog, regular dog training provides a somewhat similar opportunity.

    You need to establish a basic relation with your dog before you start to improve his IQ. Develop a rapport with him. The dog wants to do things with you and this helps to develop the relationship. The doing can be from a passive activity to an active one. Petting, brushing and feeding are more on the passive side. Taking the dog out for a walk is a more active action, but throwing a stick in the water for him to retrieve is even more active. One of our German Shepherds, Lynn, waits for my return from work. As soon as she sees me, she finds her piece of wood which she wants me to throw in the pond for her to retrieve. When you have this kind of relationship with your dog, it is easier to teach him and raise his IQ.

ACTIVITIES AND EXPERIENCES

Activities or experiences to improve a dog's intelligence fall into three categories:

■ There needs to be an availability of a variety of toys which the dog can manipulate, play with or tear up at will. These include hard rubber balls, metal pans (some with and some without handles), heavy cardboard boxes, pieces of heavy rubber (not inner tubes), small tires, sturdy plastic bottles which have been thoroughly cleaned inside and do not contain anything toxic, pieces of wood, etc. In selecting items to give a dog, one must be sure they are not objects that can be chewed on and swallowed or that have sharp edges that might cut the dog. In Germany people often play with a dog by having him fetch stones. Some of these dogs have their teeth badly worn and even broken. It is not a recommended practice.

One of our friends who has a kennel provides an enriched environment for her dogs by giving them a variety of toys. These include balls of appropriate size so that the dogs cannot swallow them, rawhide chews, tin cans with smooth and not sharp edges, heavy socks tied in a knot, old dog leads or pieces of rope on which they can pull, metal pans and pots with or without handles which they can carry around, throw on the ground and make noises with, etc. A radio or tape recorder plays much of the time to provide auditory inputs and to soothe the dogs.

■ An exposure to as wide a range of experiences as possible is recommended. These might include some tick and striking clocks, radio, television, various kitchen and other household sounds, traffic rushing along the street (with the dog on a short lead and carefully protected) and walking through a busy shopping center.

Our dog Viking, a 2-year-old German Shepherd male, almost insisted on going in the car with my wife. Since she was planning only very short stops, my wife let him go because he did not have to stay in a parked car for more than a few minutes at a time. One stop was at the country store that sells feed, seeds, insecticides, etc. Margaret had heard that this store had big scales where large dogs could be weighed. After making her purchase she asked the clerk about this and said that she had a German Shepherd in the car and could she bring him in. This was Viking's first experience in a store. He sniffed the many new smells, looked at the strange arrangement of shelves, bags, pots and other articles in the room and was led around to the scales. At first he was reluctant to get on. The platform on the scales was in fact a bit small for a dog of his size. He needed some coaxing to get him on them in a certain position. On the next trip weighing him will be much easier.

Dogs learn a great deal by imitation. In our kennel we use automatic waterers or *lixits*. We have observed that puppies as young as 6 weeks old learn to use the lixits by imitating their elders. Often at a very early age, they learn through imitation to paw a screen door ane pull it toward them in order to open it. Earlier than this they can, of course, push a screen door open and go out.

■ Formal training is important. Numerous books have been written and various training methods devised to train dogs, both to quality for AKC Obedience Degrees and to prepare them, particularly the working dogs, for such functions as protection or for guidance for the blind. Methods of formal training are beyond the scope of this book, but it is unquestionably true that the dog's IQ is raised substantially following such training. In the opinion of these authors the training, wherever possible, should be done by the dog's owner in a training class run by qualified trainers in a dog course in their area. It is also important for the owner in all contact with the dog to be consistent in his commands. If the owner does this, the dog quickly learns to associate an act with a particular word.

SUPPLEMENTS TO FORMAL TRAINING

There are various things that one can do to supplement formal training. The five mental functions already covered are receptive verbal knowledge, common sense and judgement, concentration ability, attention and problem-solving ability. These are related to some extent to formal training. For example, in receptive language the following are some of the words which a dog must learn in order to pass basic obedience required for The American Kennel Club's Companion Dog degree. These are *heel, sit, stay, wait, come* and *down*. Various other words can be easily taught to the dog in connection with particular situations. For example, our dogs learn rather early the word, downstairs. When they hear it, they go out of the house and down a flight of stairs into the yard.

We are recommending the use of conditioning principles to improve your dog's IQ. Other training methods can be used. The conditioning principles use positive reinforcements or rewards. They can be a powerful tool in training, shaping and maintaining desirable behavior. By shaping behavior we mean changing it in character, in strength or vigor, or lengthening the time it can be carried out. We can shape behavior patterns so that they occur on command. We can shape the behavior pattern of sitting so that the dog sits when we tell him and where we tell him. We can also lengthen the pattern so that he stays sitting until we release him.

To motivate our dog we have found that one of the best methods is through hunger. Do not be concerned about starving your dog. A dog can go without food for days, although he must have water. For a good training session have your dog comfortably hungry. In the opinion of Leon F. Whitney, D.V.M. the dog will train most effectively when he is about 36-hours hungry. This means that you train every other dog? During the training period many good tasting tidbits should be used as rewards or reinforcers. At the end of the training session give the dog the remainder of his meal.

REINFORCEMENTS

What are rewards or positive reinforcements? In general, rewards are anything that is pleasant to a dog. Food of the sort that the dog likes is an example of a reward. It is important that you select food that your dog likes and use it in your training when he is hungry. Social rewards can also be effective reinforcers. This includes anything that has to do with the pleasure a dog can get from a person. Words of praise, smiles and stroking the dog under the chin are excellent social rewards. The usefulness of this social kind of reward can be increased by combining it with other kinds of rewards such as food.

Repeat only behavior patterns which you want the dog to repeat later on. Generally reinforce only behavior which occurs on command. If you say, "Sit" and the dog sits, reward him. If he has already been trained to sit and he does so without a command, do not reward him. If you are having difficulty teaching the dog to sit, and he suddenly sits without your request, immediately say, "Sit, good boy" and give him a tidbit as fast as you can. Rewards should be given as soon as possible after the dog does the desired act. If you are not quick in rewarding him, he will not associate what you want him to do with the reward. If you are praising the dog, use a clear, pleasant tone of voice and simple words. Give commands each time using the same words, same tone of voice and the same gestures. Let your commands sink in. All the persons who work with a particular dog should use the same words and same gestures in giving commands.

If a dog begins to get negative and does not respond adequately, you may be pushing him too hard. Ask yourself, "Am I teaching him too many things? Is he bored? Are there other competing activities? Am I asking him to learn too fast?" Training should be fun for the dog.

Test Number 10, Receptive language (Vocabulary) has three items: responding to his name, sit and come. Your dog can increase his IQ by increasing the size of his receptive language. It is known that some dogs can learn at least as many as 400 words. You build up a vocabulary by always using the same word in connection with a particular object or action. For example, some of our dogs have learned the word, paper. They learned this word from fetching the newspaper that is thrown at the entrance to the property. It is well to keep words as simple as possible. For example, say "paper" rather than "newspaper."

If your dog does not know how to sit, then this might be the place to start your language development program through behavior modification procedures. Your dog should be comfortably hungry. Get a tidbit that he likes and show it to him. In many cases he will sit down hoping that you will give it to him. If he sits, say "Sit" and at the same time praise him and give him the tidbit. Do this again. He will soon learn that the praise, food and the word, "sit" go together. We are now beginning to condition him. It is important that you give him the tidbit immediately after he sits and as you say, "Sit". Sometimes your dog may not sit as you move toward him with a tidbit. If he doesn't, move away from him in order to draw him forward and repeat showing him the tidbit. Urge him to take it as he sits. When he sits, say "Sit", praise him and give him the tidbit. After several trials of this kind, he will begin to sit on command without reinforcements. Repeat this exercise for several days until your dog is thoroughly trained.

Common sense and judgment seem to be developed in large part through exposure to a wide variety of experiences. For example, when we lived in the country we had a dog that learned entirely on his own how to catch and kill snakes without ever being bitten. On the sea coast when a dog is confronted with a crab that has crawled up on land, it takes no more than one experience to learn to keep away from the pincer claws. Many dogs that are determined to run loose are killed by passing cars. Some that are fortunate enough to experience a near-hit survive and learn on their own the danger of cars. This aspect of common sense or of survival skill can be taught to an extent by walking the dog on a leash along a busy highway and pulling him back with a sharp "No". In addition to such survival skills, experiences similar to those in some of the tests, such as food behind a barrier, can be taught. If you dog sleeps in a dog crate, as many do, it should be quite easy to teach him to open the unlatched door of a crate whether it opens toward him or not. A crate that we keep in the house and use from time to time to confine a dog generally has the door unlatched. The door swings open and the dog must catch his paw in the door in order to pull it out and get into the crate. By placing food in the crate when the dog is outside of it, he should be able to learn this skill quite easily.

Almost any kind of training will tend to develop skills and raise the dog's IQ. To develop any skill you must first get the dog to concentrate. There are a large number of problem-solving situations. These involve concentration to some extent. Test Number 11, an alternate item, is a measure of the dog's problem-solving ability. A certain amount of concentration is necessary to do well. Suppose that your dog does poorly on this test and you want to teach him to do better. By teaching him this item he will improve this problem-solving ability and increase his concentration.

Behavior modification can help your dog solve this problem. How is this done? Put your dog on a leash and attach the leash to some object like the bumper of your car or fence. Hold the glass in front of and within the reach of the dog. Place a tidbit along side of the glass with the dog looking on and urge him to get the food. If he gets the goody, then place another tidbit about 1 inch from the edge and behind the glass. Urge him to get the food. If he succeeds, place the next tidbit more toward the middle of, and behind the glass, as shown in the test itself (Chapter 9). If he fails, bring the food out closer to the edge so that it will be easier for him to solve the problem. You may be able to think of another problem-solving experience for the dog to work on which will improve both his problem-solving ability and concentration.

To improve your dog's attention, expose him to different kinds of noise other than those mentioned in Test Number 7. Take your dog out in a fenced area and let him run loose. Find two flat pieces of wood and hide them behind your back. As your dog is attending to something else, slap the wood together loud enough for the dog to hear. If he does not become attentive and move toward you and the pieces of wood, repeat the slapping. Other objects that can be used include a small transistor radio (increase the volume gradually), two tennis ball cans, and horn or bell on a bicycle.

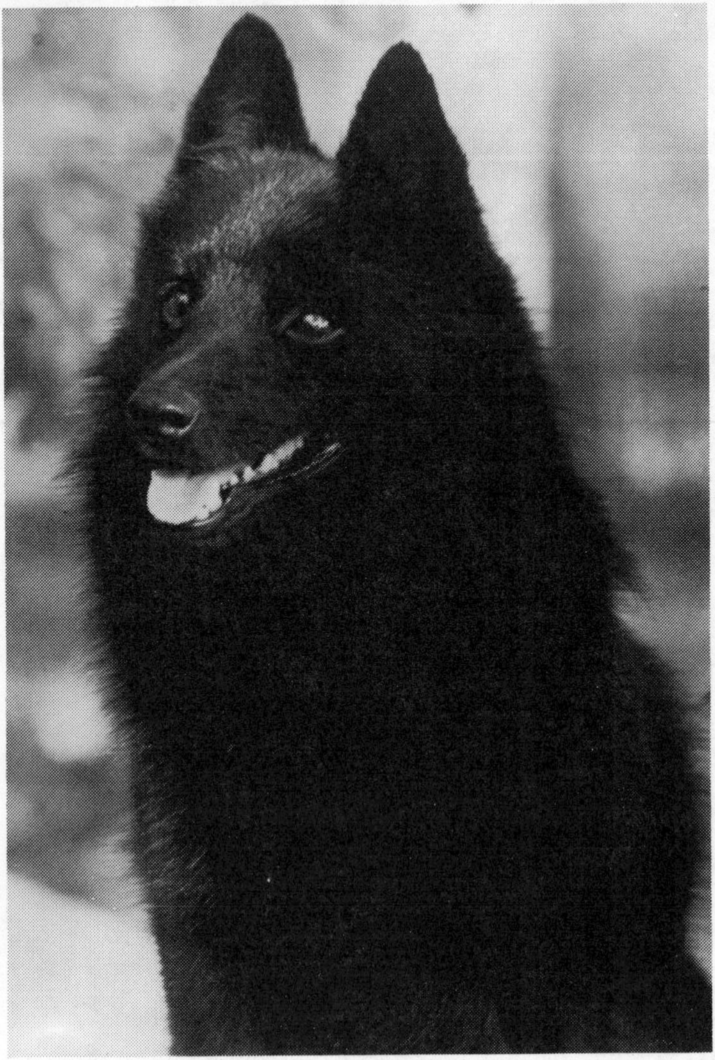

Chapter 11
Differences in Intelligence Among Breeds

Judging how intelligent a dog is in comparison to another dog is difficult, according to the Encyclopedia Britannica (Macropeadia). This summary of knowledge adds that intelligence may be measured by the ease with which a dog can be trained. This may depend in large measure, however, on the degree of motivation. When individual dogs are sufficiently motivated there appear to be no wide differences in intelligence between breeds. Certain breeds, particularly the shepherd dogs and poodles have the reputation for high intelligence.

Various breeds have been selected for various uses. Hunting breeds, such as hounds, have been selected to work independently of human handlers and are somewhat difficult to keep under restraint. Shepherd dogs, on the other hand, have been selected for their ability to learn to work under direction and to form firm habits. It is easier to teach them restraint than the hunting breeds.

BREED RANKS

"The German Shepherd must be ranked among the three or four most intelligent purebred dogs in the world," according to Jones and Rendel. They add, "The poodle would also have to be included, and one or two others whose selection must be left to the impartial judgment of dog fanciers everywhere. He (the German Shepherd) has a remarkable record as the principal guide dog for blind persons. He is, after experience in two World Wars, the only breed accepted for military service by the United States Armed Forces. He has performed outstandingly in obedience competition. The poodle has always been an intelligent and attractive dog. But recently he has been rediscovered and has become one of the most popular purebred dogs in the nation. He has the beauty, brains and unsurpassed style to be a desirable pet and an effective show dog. He is a star of obedience competition and, if need be, a rugged field dog, although his uses as a hunter have dwindled."

Jones and Rendel also include the Labrador Retriever, Golden Retriever and the Doberman Pinscher among the ten most intelligent dogs. They say that there is not a more intelligent dog than the Labrador Retriever. About the Golden Retriever they say that his intelligence makes him easy to train, that he scores consistently in obedience training and that some Goldens are guide dogs for the blind. They describe the Doberman Pinscher as keenly intelligent, high-spirited and eager to follow his master's bidding.

Coon used her test to examine intellectually 35 different breeds of dogs. She found that among different dog groups, the non-sporting dogs have the highest IQ. They are followed in turn by hounds, sporting dogs, working dogs, terriers and toys. She also determined that the smartest dog was the Standard Poodle. The next 10 smartest dogs in order, according to Coon are as follows:

- Bulldog
- Beagle
- English Springer Spaniel
- Golden Retriever
- Basset Hound
- (Smooth-coated)Dachshund
- Great Dane
- Collie
- Belgian Tervuren and Australian Shepherd (sic). These two tied.

The following are next in Coon's rank:

- Miniature Poodle
- German Shepherd, Pionter (Tied)
- Airedale Terrier, Toy Poodle, Silky Terrier (Tied)
- Schipperke, Cocker Spaniel, St. Bernard (Tied)
- Labrador Retriever, Miniature Schnauzer (tied)

- Boxer
- Pekingese
- Whippet
- Shetland Sheepdog
- Doberman Pinscher, Lhasa Apso (Tied)
- Irish Setter
- Pug
- West Highland White Terrier
- Afghan Hound
- Siberian Husky
- Pomeranian
- (Smooth-coated) Chihuahua

RESULTS

The most stupid dog was the (smooth-coated) Chihauhua. These dogs are ranked 35th. By Coon's scores the Standard Poodle is four times as smart as the Chihuahua. From the authors' observations it is doubtful that the Chihuahua is all that stupid. Some experienced and knowledgeable dog persons think that the Chihuahua is intelligent.

Jones and Rendel say this about the Chihuahua, "It is the smallest dog in the world. He may weigh as little as a single pound. The breed excells in obedience training. Properly trained, the bright, little Chihuahua takes a back seat to none. Besides intelligent and dimunitive size, the breed has appeal because of its variety of coats and colors."

One questions the validity and usefulness of the data presented by Coon. No information is given as to the number of dogs tested in each category or the ages of the dogs. She herself said, "These norms will not give you a definite answer regarding how smart a given breed is since small numbers of dogs were used for the comparison samples."

In the development of norms for human intelligence, such as the Wechsler Intelligence Scale for Children—Revised (WISC-R), the standardization sample included 200 children in each of eleven age groups, ranging from 6½ through 16½ years. The total sample contained 2200 cases. At each age level the sample included 100 boys and 100 girls. Only when studies are made on a relatively large population can any confidence be placed on the relevant standing of individuals (dogs, humans or other animals).

Coon gives no information as to the age of the dogs she tested. Age obviously makes a big difference. Age is also an important factor in determining intelligence in humans. For example, in the Digit Span subtest of the WISC-R we would expect a 16-year-old with an average IQ (between 90 and 109) to be able to repeat after the examiner about twice as many digits forward and backward as would be expected of a 6 year old of the same intelligence. This age factor alone would completely invalidate any of Coon's IQ ratings of dogs of different breeds without taking into account age.

The German Shepherd is tied for 13th place with the Pointer, and the Doberman Pinscher is tied for the 27th place with the Lhasa Apso. Yet, the Shepherd and the Doberman are used to a great extent as working dogs for all sorts of complex tasks: search and rescue, guide dogs, war dogs, etc. How could this be if these two dogs are no more intelligent than Coon's test indicates? What does her test really measure?

Coon writes that there is no difference in IQ between non-kennel and kennel German Shepherds. She also reports that there is no difference in IQ between German Shepherds who live in the house and those who live outside. She implies that there is no difference between house pets and kennel dogs. This seems most unlikely because house pets are generally exposed to a much wider range of experiences than are kennel dogs. One might compare the human situation in which it has been found that children's IQs can be significantly raised by placing them in an enriched environment.

Index

A
Ability to adjust to motivation	66
new situation	66
Ability to detect movement	61
form concepts	53
remember	51
Ability to see things, as missing	65
as wrong	65
Abstract	36
Abstraction ability	65
Abstraction of triangularity	54
Activities, to improve dog's IQ	147
Age of cutting	99
Animals, testing of others	83
Apprehension	8
Arithmetic	34
Attention	60
Avoidance	48

B
Basenji, experiments	95
Beach, F. A.	85
Beagle, experiments	95
Binet, Alfred	15
Bingham, W. E.	86
Biological needs	79
Block design	39
Breed ranks	155

C
Cattell, James McKeen	24
R. B.	20
Clarke, R. S.	85
Cocker Spaniel, experiments	95
Coding	40
Common sense	60
Comprehension	33
Concentration ability	62
Concept of persistence of behavior	19
Conceptual	36
Conceptualization, 3 levels	36
Conceptual thinking	41
Concrete	36
CR	48
Critical periods	97
7 to 12 weeks	99
3 to 7 weeks	98
12 to 16 weeks	99
up to 3 weeks	97
Crystallized intelligence	19
CS	48

D
Deprivation of perceptual experiences, research of	85
Digit span	33
Digit symbol	40
Distractions	62
DIQ, obtaining	109
Dog intelligence, critical periods	97
quotient	109
Dominant	94
Ducklings, study of	82

E
Early deprivation studies	90
Environment, effects of	89
Escape	48
Examiner	104
Experiences, to improve dog's IQ	147
Eye-muscle coordination, fine	65

F
Fluid intelligence	19
Formal training, supplements	149
Form L-M, mental age	26
scoring	26
Free-environment study	71
Functional	36

G
Griffiths, W. J.	86
Guilford, J.P.	20

H
Hebb, D. O.	84
Human infants vs. puppies	69
Hymovitch, Bernard	84

I
Inference	51
Inherited characteristics	93
classifications	94
IQ	27

Insightful learning	46
Insight learning	48
Instrumental conditioning	48
Intellectual functions	14
Intelligence	13
crystallized	19
fluid	19
scales	17
synonyms	13
tests	15

J

Judgment	60

L

L & L Dog IQ Test	57
Language	41
Learning	46
insight	48
insightful	46

M

Memory	16, 41
short-term	52
Mental age	16, 26
Mental test	24
Mothering, deprivation of	75
Multiple factor analysis	16

N

Non-intellective factors	19
Number	16
Numerical reasoning	42
Nygar, Ronald	19

O

Object assembly	39
Operant conditioning	48
Overpampering, by owners	86

P

Pavlov, Ivan	48
Perceptual ability	17
Performance Scale	32
Performance subtests	37
age—intelligence scale	40
block design	39
coding	40
digit symbol	40
mazes	40
object assembly	39
picture arrangement	37
picture completion	38
Persistence to intelligence, relationship	19
Pfaffenberger, Clarence	96
Piaget, J.	20
Picture arrangement	37
Picture completion	38
Picture vocabulary	47
Planning ability	65
Polygenic character	94
Primary mental abilities	16
Problem box	50
Problem-solving ability	50, 66

R

Rats, study of	84, 85
Reality testing	33
Reasoning	16, 41, 51
numerical	42
Receptive language	46, 59
Recessive	94
Reinforcements	150
Response, conditioned	48
unconditioned	48
Restrictive surroundings	91
Riesen, A. H.	83
Riess, B. F.	85

S

Scoring	107
Scotties	85
Sex-linked character	94
Shetland sheep dog, experiments	96
Short-term memory	52
Skinner box	48
Social intelligence	38, 42
Space perception	65
Spatial	16
Spearman, Charles	16
Squeeze boxes	86
Standardized tests	23
Stanford-Binet Test	25
Stanford Revision of the Binet Scale	24
Staring syndrome	105
Stimulus, conditioned	48
unconditioned	48

T

Terman, Lewis M.	24
Test preparation	106
results	108
sequence	106
Tests	113-145
Time limit	107
Thorndike, E. L.	16
Thurstone, L. L.	16

U

UCR	48
UCS	48

V

Verbal	16
Verbal knowledge, fund of	57
Verbal Scale	32
Verbal subtests	32
arithmetic	34
comprehension	33
general information	32
similarities	35
digit span	33
vocabulary	37
Visual motor	42
Vocabulary	37

W

WAIS	17, 29
subtests	32
Wechsler Scale	28
Adult Intelligence Scale	17, 29
Intelligence Scale for Children	17, 28
Wechsler	28
Scales, underlying theory	31
Wechsler, David	17
WISC	17, 28
subtests	32
Wire fox terrier, experiments	96
Word fluency	16